Green Energy and Technology

Climate change, environmental impact and the limited natural resources urge scientific research and novel technical solutions. The monograph series Green Energy and Technology serves as a publishing platform for scientific and technological approaches to "green"—i.e. environmentally friendly and sustainable—technologies. While a focus lies on energy and power supply, it also covers "green" solutions in industrial engineering and engineering design. Green Energy and Technology addresses researchers, advanced students, technical consultants as well as decision makers in industries and politics. Hence, the level of presentation spans from instructional to highly technical.

Indexed in Scopus.

More information about this series at http://www.springer.com/series/8059

Wilson Acchar · Vamberto Monteiro da Silva
Editors

Use of Cassava Wastewater and Scheelite Residues in Ceramic Formulations

Springer

Editors
Wilson Acchar
Department of Physics
Federal University of Rio Grande do Norte
Natal, Rio Grande do Norte, Brazil

Vamberto Monteiro da Silva
Science and Technology of Paraíba
Federal Institute of Education
João Pessoa, Paraíba, Brazil

ISSN 1865-3529 ISSN 1865-3537 (electronic)
Green Energy and Technology
ISBN 978-3-030-58781-9 ISBN 978-3-030-58782-6 (eBook)
https://doi.org/10.1007/978-3-030-58782-6

This Springer imprint is published by the registered company Springer Nature Switzerland AG
The registered company address is: Gewerbestrasse 11, 6330 Cham, Switzerland

Contents

Cassava Wastewater: An Introduction, Characterization and Potential

Jônatas Macêdo de Souza

Abstract This chapter contains a brief introduction on the need to reuse waste, presenting cassava wastewater as an effluent from producing flour and starch, which must be properly disposed of. The characteristics of cassava wastewater as well as its chemical composition and potential uses for reuse are discussed.

Keywords Agricultural waste · Ceramic formulations · Reuse

The growing exploitation of natural resources is becoming increasingly worrying due to the environmental impacts caused by extracting raw materials and the consequent inappropriate disposal of products at the end of their useful life. In view of this, the number of studies aimed at minimizing the effects resulting from degrading nature caused by various sectors of the economy has increased in recent years. A major challenge faced by researchers is to develop new, sustainable materials which reduce damage to ecosystems while meeting the requirements required by technical standards and environmental legislation.

It is interesting to differentiate between solid residues and wastes for a better understanding of the subject. In Brazil, Law No. 12,305/10 [1] defines solid residue as: "material, substance, object or discarded good resulting from human activities in society, whose final destination is proceeded, proposed to proceed or is obliged to proceed in the solid or semisolid states, as well as gases contained in containers and liquids whose particularities make its release in the public sewerage network or in bodies of water unfeasible, or require technical or economically unviable solutions in view of the best available technology." Furthermore, waste is defined as: "solid residues which, after all the possibilities of treatment and recovery by available and economically viable technological processes have been exhausted, do not present any possibility other than final and environmentally appropriate disposal."

J. M. de Souza (✉)
Graduate Program in Materials Science and Engineering (PPGCEM), Technology Center (CT), Federal University of Rio Grande do Norte (UFRN), University Campus, Natal, Rio Grande do Norte, Brazil
e-mail: jonatasms2@ufrn.edu.br; jonatasms2@hotmail.com

W. Acchar and V. Monteiro da Silva (eds.), *Use of Cassava Wastewater and Scheelite Residues in Ceramic Formulations*, Green Energy and Technology,
https://doi.org/10.1007/978-3-030-58782-6_1

The concern with the correct destination of waste together with the problem of scarcity of raw material has stimulated several works with the use of these environmental liabilities in manufacturing ceramic materials; for example, Azevedo et al. [2] studied the effect of the glass residue on the rheological properties of the adhesive mortar, noting that mixtures with 10% replacement of cement and 15 and 20% replacement of aggregate, both concerning mass, presented better rheological characteristics; Nascimento et al. [3] incorporated granite cutting waste in plastering mortar and found that the mortar's mechanical properties were improved when the residue replaced the sand in the contents of 5, 10 and 20%; Spósito et al. [4] replaced part of the sand with PET bottle waste in the manufacture of hydrated Portland cement-based mortars and stated that this might be another possibility for disposal of the waste.

In other studies, Meena and Luhar [5] researched concrete produced with treated wastewater instead of potable water and found a compressive strength between 85 and 94% of the reference concrete; Yang et al. [6] analyzed the properties of foam concrete containing brick powder derived from construction and demolition waste and found that the residue could replace cement by weight by up to 15%; Siqueira et al. [7] verified the behavior of soil-cement bricks with eggshell and slag wastes and observed that the eggshell waste could replace mass cement by up to 30% and the slag residue could replace mass soil up to 15%; Islam et al. [8] used fly ash combined with cement to manufacture compacted earth blocks and verified that 7–8% cement and 15–20% fly ash concerning on the weight of the dry soil provide compressive strength and durability in terms of water absorption according to the standards of England and Australia. The results obtained in these studies of incorporating residues into ceramic formulations demonstrate that these mixtures are presented as alternative solutions for the correct destination of various environmental liabilities, as long as the wastes do not impair the technical properties required for using the materials.

Despite an extreme relevance, many economic sectors generate waste during the production process which are harmful to the environment when improperly disposed. This is the case in the agricultural sector in which the processing of some crops presents organic waste which needs further studies to determine the correct destination or reuse of this waste. One of these cultivars is cassava (*Manihot esculenta Crantz*), which has worldwide production of around 285 million tons/year of unprocessed roots, being the fifth most important basic crop in the world [9]. According to the Brazilian Institute of Geography and Statistics [10], Brazilian production of cassava root was around 19.4 million tons in 2018, which places Brazil as the fourth largest world producer behind only Nigeria, Thailand and Indonesia.

Cassava plays an important role in the social and economic spheres in family farming. The agricultural product is often the main crop produced by certain regions. Thus, many families work in the cultivation, processing and commercialization of the generated cassava products. The roots are usually processed in industries with little technology, giving rise to by-products.

In the production of flour, are generated two types of wastewater, the first type is derived from the water used to wash the roots, and the second is the liquid extracted from the roots during pressing [11] which is commonly called "*Manipueira*" in

Fig. 1 Cassava wastewater storage tank

Brazil, a yellowish liquid with a high concentration of organic matter, nutrients and hydrocyanic acid. In the production of flour, the effluent (*Manipueira*) is disposed of separately immediately after pressing, while in the production of starch, it is mixed with the high volume of drinking water used in the various stages of crushing, sieving and sedimentation. Because of this, the volume of wastewater generated in the starch production process is higher than that verified in the production of flour.

The estimated generation of *Manipueira* in flour houses is about 300–600 L for each ton of processed cassava roots [12, 13]. Meanwhile, in the starch factories, which have high technology, according to the literature, 2.75–6.3 m^3 of cassava wastewater (*Manipueira* and water) are generated for each ton of roots processed [14, 15]. According to the Food and Agriculture Organization of the United Station [11], this volume may be even higher, depending on the process used and the country, the amount of wastewater per ton of roots processed in the starch industries can reach up to 18 m^3.

In Brazil, most of the cassava production is for the manufacture of flour. The estimate is that the product consumes 60–80% of the total production of cassava roots [16–18]. Trying to provide more alternatives for the destination of the generated effluent, in this book, were conducted several kinds of researches with the use of *Manipueira* from flour houses.

From now on, whenever the term "cassava wastewater" is mentioned in this book, it is understood that it is the "*Manipueira*" extracted in the production of flour. Figure 1 shows the cassava wastewater stored in a tank right after processing cassava in a flour mill.

Madeira et al. [19] claim that the toxicity of this effluent represents a major environmental problem, as it is usually not treated before being discarded. When it is considered the processing of 70% on the average Brazilian production of 22.5 million tons/year and a generation of 300 L per ton of cassava wastewater, in the flour production, approximately 4.73 billion liters of the effluent are generated annually just in Brazil. This fact demonstrates a need for alternatives for the correct destination

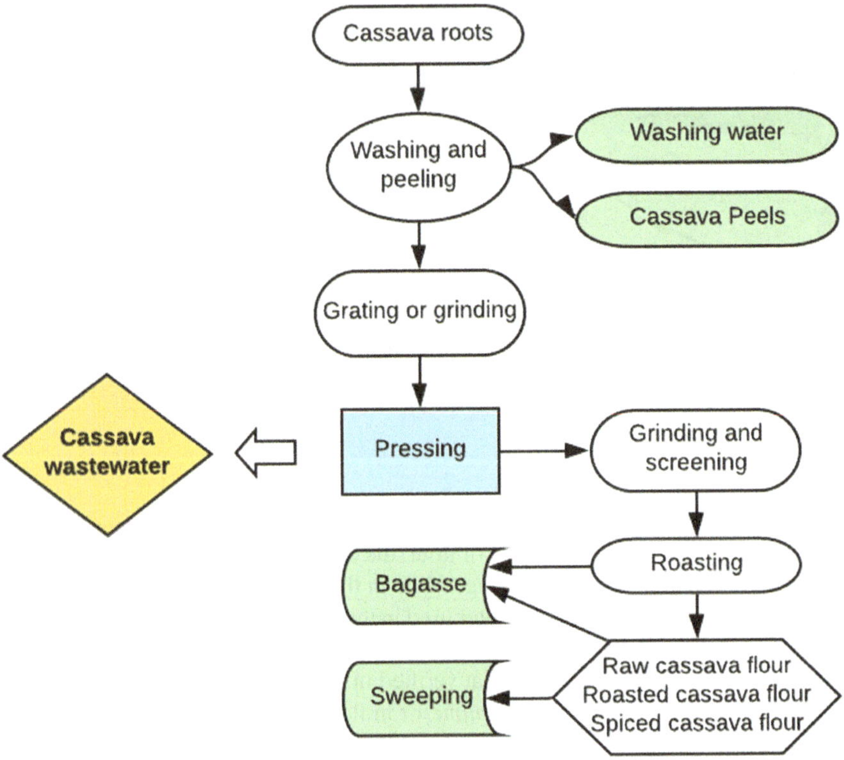

Fig. 2 Steps of the cassava flour manufacturing process

of this waste. A flowchart in Fig. 2 shows the origin of the cassava wastewater from processing the roots in a flour mill.

The cassava wastewater composition is variable as it depends on the crop and on typical characteristics of the region where the cassava is grown, such as the type of soil and the climatic factors. Table 1 below shows the results of the chemical composition of the cassava wastewater used in the studies presented in this book.

The results obtained for the cassava wastewater characterized in this work show high potassium, nitrogen, sodium, iron and calcium concentrations, which is in accordance with the values found in the literature by Silva et al. [20]. The variations of the values obtained in comparison with the literature can be justified by the fact that the composition changes according to several factors such as climate, soil, crop, among others. The low pH presented demonstrates the acidity of the wastewater, and a high value of solids present in the liquid can also be observed, showing the presence of organic matter. The high values of chemical oxygen demand (COD), 14.70–101.38 gO2/L, and biological oxygen demand (BOD), 6.21–29.20 gO2/L, presented in the literature, express the high organic load of cassava wastewater and show that the effluent is a source of pollution [20, 21].

Table 1 Chemical composition of cassava wastewater

Parameters analyzed	Values
Nitrogen (mg/L)	1121
Phosphorus (mg/L)	132
Potassium (mg/L)	1456
Calcium (mg/L)	93
Magnesium (mg/L)	219
Sodium (mg/L)	351
Zinc (mg/L)	20
Copper (mg/L)	9
Iron (mg/L)	117
Manganese (mg/L)	15
Total solids (g/L)	14.56
Suspended solids (g/L)	2.37
Chloride (mg/L)	545.40
Sulfur (mg/L)	78.53
Phosphate (mg/L)	27.07
Cyanide (mg/L)	5.62
pH	4.5

Table 2 Chemical composition of cassava wastewater powder

Oxide	SiO_2	Al_2O_3	Fe_2O_3	K_2O	CaO	SO_3	P_2O_5	Na_2O	Br
Weight (%)	2.457	2.306	15.922	32.654	25.739	3.123	3.592	9.986	4.220

Was performed an X-ray fluorescence (XRF) assay to determine the oxides present in the waste. The settleable solids present in the liquid were used in the test. The material was dried in an oven at a temperature of 60 °C for 24 h to obtain the powdered cassava wastewater. Table 2 shows the results obtained in the XRF assay. The results demonstrate that the significant potassium (K_2O), calcium (CaO), iron (Fe_2O_3) and sodium (Na_2O) levels are similar to the results presented in the reference tests performed on the liquid, as shown in Table 1.

The cassava wastewater has high levels of solute, including starch made up of polymeric molecules which have binding properties acting as a biopolymer. Starch molecules together with solids in solution can improve the plasticity of the mixture and its resistance by filling empty spaces and their effect as a binder. The plasticizer characteristic can be seen in the study by Akindahunsi and Schmidt [22], who studied the effect of cassava starch on the contraction properties of concrete. They obtained a decrease in shrinkage with the increase in the use of cassava starch by up to 2% by mass. Abd et al. [23] studied the effect of using corn starch in a concrete mixture, obtaining an improvement in workability with an increase in density and resistance to compression with the use of 1% corn starch.

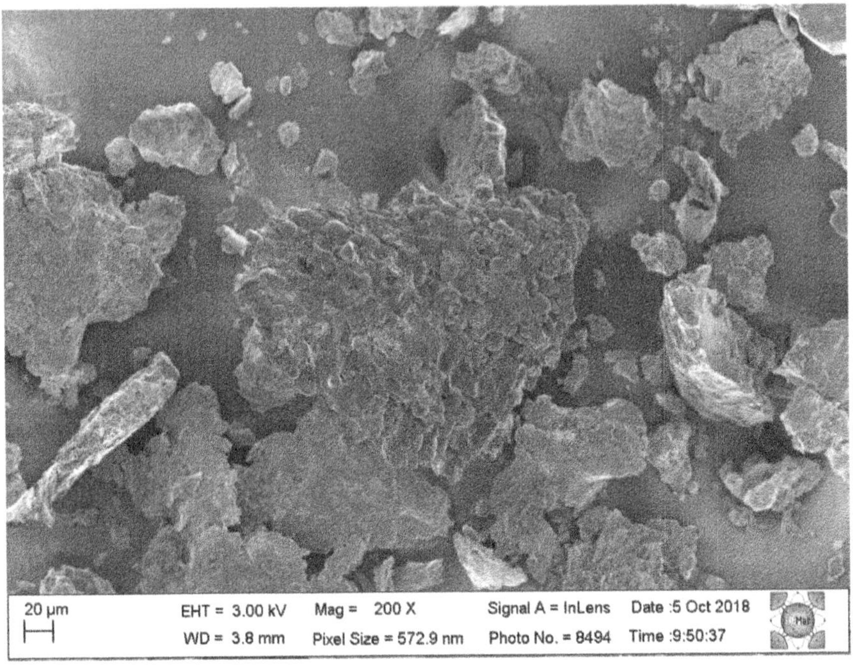

Fig. 3 SEM of cassava wastewater powder sample 200× magnification

The starch particles from the cassava starch are predominantly oval or spherical. There is also a difference in the granule size ranging from 4 to 15 μm, with an average of 10 μm for the cassava granules [24].

The powder was subjected to the scanning electron microscopy (SEM) test to observe the morphology of the solid part of cassava wastewater. The starch globules clustered can be seen in Fig. 3.

In Fig. 4, it is possible to observe oval starch globules with sizes around 10–14 μm. The globes have shapes and sizes similar to those found by Vieira et al. [24].

In relation to cyanide, there is an important concern on the part of Occupational Health agencies with the control and exposure of workers to chemical agents. The Occupational Safety and Health Administration [25] establishes a maximum occupational exposure limit to hydrocyanic acid of 10 parts per million (ppm) in an 8-h workload per day. However, the regulatory standard NR 15 [26] establishes a tolerance limit of 8 ppm for the same 8 h. Thus, cassava wastewater must be placed in an environment which allows ventilation for 48 h in order to reduce the cyanide content present in the composition and be exposed to the sun so that the hydrocyanic acid (HCN) (which has a boiling temperature of 25.6 °C) volatilizes. According to Tokarnia et al. [27] and Silva [28], cassava wastewater has its cyanide content reduced to half in 24 h by this procedure, and it further decays to less than 1/4 with 48 h. The hydrocyanic acid content shown in Table 1 was well below that found by Silva et al. [20] and Peres et al. [21], and this is due to the fact that the cassava wastewater was

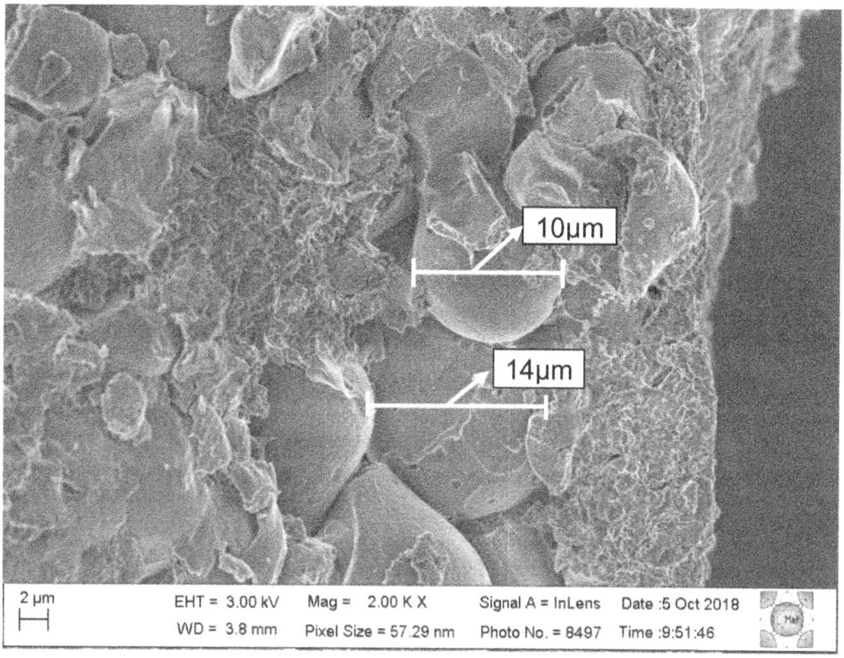

Fig. 4 SEM of cassava wastewater powder sample 2000× magnification

exposed to the sun in a ventilated environment for 2 days. It can also be added that the HCN content corresponds to 5.62 mg/L, which meets the recommendations of the Occupational Health agencies for a workload of 8 h per day.

When considering the environmental impacts caused by inadequate disposal of the cassava wastewater, it can mention some examples such as eutrophication of lakes and rivers, contamination of groundwater and the death of animals, among others. In view of this, some authors have studied ways to take advantage of the waste in order to add value to it and ensure an adequate destination. Are found several examples of uses for cassava wastewater in the literature: bio-hydrogen production [29–31]; methane production [32]; microalgae cultivation [12]; use as fertilizer [33, 34]; use as natural insecticide and pesticide [35, 36]; use in the diet of sheep [37]; production of sand blocks [38].

Araújo et al. [39, 40], for example, studied soil bricks with the substitution of water for cassava wastewater and stated that the liquid has binding properties. However, the authors did not carry out tests that prove that the effluent can be used as a stabilizer in bricks, or technical tests which demonstrate the viability of the produced material. Therefore, studies on the use of cassava wastewater in ceramic formulations must be deepened.

In everyday life, some macroscopic phenomena are often related to empirical assumptions, but these can be scientifically explained in the area of Materials Engineering by observing the microstructure of samples. Therefore, this book has the main objective to serve as a basis for students and researchers who are interested in the subject of using cassava wastewater in ceramic or composites materials. So, are presented various studies of ceramic and composites materials with cassava wastewater, where were analyzed the physical, mechanical, and microstructure properties of these materials. The following chapters will present the topics: Soil-cement bricks with cassava wastewater; ecological brick of mineral residues and cassava wastewater; mortar produced with scheelite residue and cassava wastewater; and earthbag construction system with cassava wastewater.

It is worth mentioning that the content presented in the book does not address all possible applications of cassava wastewater in ceramic materials, nor does it intend to exhaust the discussions on the subject. On the other hand, it provides a broad view of pioneering work and shows the potential for applying and utilizing agricultural waste in construction materials, thereby partially or totally replacing potable water.

References

1. Brazil (2010) Law No. 12,305, of August 2, 2010, national policy on solid waste. Union Newspaper, Brasília
2. Azevedo ARG, Alexandre J, Zanelato EB, Marvila MT (2017) Influence of incorporation of glass waste on the rheological properties of adhesive mortar. Constr Build Mater 148:359–368
3. Nascimento AS, dos Santos CP, de Melo FMC et al (2020) Production of plaster mortar with incorporation of granite cutting wastes. J Clean Prod 265:121808
4. Spósito FA, Higuti RT, Tashima MM et al (2020) Incorporation of PET wastes in rendering mortars based on portland cement/hydrated lime. J Build Eng, 101506
5. Meena K, Luhar S (2019) Effect of wastewater on properties of concrete. J Build Eng 21:106–112
6. Yang D, Liu M, Ma Z (2020) Properties of the foam concrete containing waste brick powder derived from construction and demolition waste. J Build Eng, 101509
7. Siqueira FB, Amaral MC, Bou-Issa RA, Holanda JNF (2016) Influence of industrial solid waste addition on properties of soil-cement bricks. Cerâmica 62:237–241
8. Islam MS, Elahi TE, Shahriar AR, Mumtaz N (2020) Effectiveness of fly ash and cement for compressed stabilized earth block construction. Constr Build Mater 255:119392
9. FAOSTAT (2017) Food and agriculture organization of the United Nations (FAOSTAT) statistics database. FAO, Rome, Italy. http://faostat3.fao.org/home/E. Accessed 9 Dec 2019
10. IBGE (2019) Brazilian Institute of Geography and Statistics—Monthly survey of forecast and monitoring of agricultural crops in the calendar year. Systematic survey of agricultural production, pp 1–138
11. FAO (2001) Strategic environmental assessment. An assessment of the impact of cassava production and processing on the environment and biodiversity. In: Proceedings of the validation forum on the global cassava development strategy, vol 5, Rome
12. Carvalho JC, Borghetti IA, Cartas LC et al (2018) Biorefinery integration of microalgae production into cassava processing industry: Potential and perspectives. Bioresour Technol 247:1165–1172
13. Ribas MMF, Cereda MP, Villas Bôas RL (2010) Use of cassava wastewater treated anaerobically with alkaline agents as fertilizer for maize (Zea mays L.). Brazilian Arch Biol Technol 53:55–62
14. Hasanudin U, Kustyawati M, Iryani D et al (2019) Estimation of energy and organic fertilizer generation from small scale tapioca industrial waste. IOP Conf Ser Earth Environ Sci 230:12084
15. Phu TV, Thao NT, Son CK et al (2016) Potential use of by-products for animal feed from cassava in Vietnam. J Sci 14:93–100

16. Mombo S, Dumat C, Shahid M, Schreck E (2016) A socio-scientific analysis of the environmental and health benefits as well as potential risks of cassava production and consumption. Environ Sci Pollut Res 24:5207–5221
17. Okudoh V, Trois C, Workneh T, Schmidt S (2014) The potential of cassava biomass and applicable technologies for sustainable biogas production in South Africa: a review. Renew Sustain Energy Rev 39:1035–1052
18. de Oliveira Chaves YA, Val Springer M, Boloy RAM et al (2020) Performance study of a microturbine system for cogeneration application using biogas from manipueira. BioEnergy Res 13:659–667
19. Madeira JGF, Boloy RAM, Delgado ARS et al (2017) Ecological analysis of hydrogen production via biogas steam reforming from cassava flour processing wastewater. J Clean Prod 62:709–716
20. Silva FF, Bertonha A, Freitas PSL et al (2005) Fragment of starch to glucose to identify waste water from manioc industry in depth of soil. Acta Sci Biol Sci 27:507–511
21. Peres S, Monteiro MR, Ferreira ML et al (2019) Anaerobic digestion process for the production of biogas from cassava and sewage treatment plant sludge in Brazil. BioEnergy Res 12:150–157
22. Akindahunsi AA, Schmidt W (2019) Effect of cassava starch on shrinkage characteristics of concrete. African J Sci Technol Innov Dev 11:441–447
23. Abd SM, Hamood QY, Khamees AS, Ali ZH (2016) Effect of using corn starch as concrete admixture. Int J Eng Res Sci Technol 5:35–44
24. Vieira JC, Montenegro FM, Lopes AS, da Pena RS (2010) Influence of the addition of cassava starch on the characteristics of tea-type bread. Bol Cent Pesqui Process Aliment 28:37–48
25. OSHA (2018) Permissible exposure limits/OSHA annotated table Z-1. https://www.osha.gov/dsg/annotated-pels/tablez-1.html. Accessed 18 Dec 2019
26. Brazil (1978) Ministry of labour. NR-15-unhealthy activities and operations - annex 11
27. Tokarnia CH, Brito MF, Barbosa JD et al (2012) Toxic plants of Brazil for food animals. Helianthus, Rio de Janeiro (Brazil)
28. Silva VC (2016) Experimental poisoning by cassava the by-product of Manihot esculenta Crantz (manipueira) in sheep. Master dissertation. Federal University of Bahia. Salvador/BA, 75 p
29. Andreani CL, Tonello TU, Mari AG et al (2019) Impact of operational conditions on development of the hydrogen-producing microbial consortium in an AnSBBR from cassava wastewater rich in lactic acid. Int J Hydrogen Energy 44:1474–1482
30. Meier TRW, Cremonez PA, Maniglia TC et al (2020) Production of biohydrogen by an anaerobic digestion process using the residual glycerol from biodiesel production as additive to cassava wastewater. J Clean Prod 258:120833
31. Wadjeam P, Reungsang A, Imai T, Plangklang P (2019) Co-digestion of cassava starch wastewater with buffalo dung for bio-hydrogen production. Int J Hydrogen Energy 44:14694–14706
32. Chavadej S, Wangmor T, Maitriwong K et al (2019) Separate production of hydrogen and methane from cassava wastewater with added cassava residue under a thermophilic temperature in relation to digestibility. J Biotechnol 291:61–71
33. De Araújo NC, De Lima VLA, de Lima GS et al (2019) Nutrient contents and growth of corn fertigated with human urine and cassava wastewater. Rev Bras Eng Agrícola e Ambient 23:681–686
34. de Araújo NC, Lima VLA, Ramos JG et al (2019) Contents of macronutrients and growth of "BRS Marataoã" cowpea fertigated with yellow water and cassava wastewater. Rev Ambient Água 14(3):e2309
35. De Carvalho JD, Neves FL, da Silva CD, Bittencourt MAL (2017) Biological aspects and insecticide action of plant species on eggs and nymphs of citrus black fly (Aleurocanthus woglumi Ashby - Aleyrodidae) at laboratory level. Rev. Bras. Frutic. 39, n.5: (e-045)
36. Zevallos DMP, Pereira Querol M, Ambrogi BG (2018) Cassava wastewater as a natural pesticide: current knowledge and challenges for broader utilisation. Ann Appl Biol 173:191–201
37. Santana Neto JA, Martinele I, Cedrola F et al (2017) Apparent digestibility and rumen protozoal profile of sheep fed cassava wastewater. Rev Bras Saúde e Produção Anim 18:327–336

38. Akpokodje IO, Uguru H (2019) Compressive strength of commercially produced sandcrete blocks within isoko metropolis of delta state, Nigeria. Turkish J Agric Eng Res 1:91–103
39. Araújo NC, Queiroz AJP, Ramos AP et al (2015) Thermal degradation of brick made of soil and cassava wastewater. Electron J Manag Educ Environ Technol 19:861–870
40. Araújo NC, Ramos AP, Queiroz AJP, dos Santos RC (2015) Mechanical properties of manufactured bricks with soil and cassava wastewater. Rev Verde 10:213–217

Soil-Cement Brick with Cassava Wastewater

Jônatas Macêdo de Souza and Luciana de Figueiredo Lopes Lucena

Abstract This chapter covers the use of cassava wastewater in manufacturing soil-cement bricks. Physical, mechanical and analysis tests on the microstructure of the samples were performed to verify the technical feasibility of substituting potable water for cassava wastewater in bricks. The results obtained for bricks with cassava wastewater are compared to bricks which did not use cassava wastewater in their composition, and also with the technical standard requirements.

Keywords Ecological bricks · Physical properties · Mechanical properties · Microstructural properties

1 Introduction

The soil has been used as a building material in civil engineering works since ancient times. Seco et al. [1] claim that the worldwide technique of building structures using soil-based materials has been used for more than 9000 years and is still very present in developing countries. The success of this material is linked to a large amount of available raw material, simplicity in the executing the technique, economy, excellent thermal and acoustic properties [2–4], and the ease of reuse with grinding and wetting. However, the works built with raw soil have a high susceptibility to water as a principal problem [5, 6].

J. M. de Souza (✉)
Graduate Program in Materials Science and Engineering (PPGCEM), Technology Center (CT), Federal University of Rio Grande do Norte (UFRN), University Campus, Natal, Rio Grande do Norte, Brazil
e-mail: jonatasms2@ufrn.edu.br; jonatasms2@hotmail.com

L. de Figueiredo Lopes Lucena
School of Science and Technology (ECT), Federal University of Rio Grande do Norte (UFRN), University Campus, Natal, Rio Grande do Norte, Brazil
e-mail: luciana@ect.ufrn.br

W. Acchar and V. Monteiro da Silva (eds.), *Use of Cassava Wastewater and Scheelite Residues in Ceramic Formulations*, Green Energy and Technology,
https://doi.org/10.1007/978-3-030-58782-6_2

The soil-cement technique appears as one of the solutions for water susceptibility, since the addition of a binding agent (in appropriate minimum levels) coupled with high compaction energy provides soil stabilization. One of the materials produced from this technique and commonly used in house construction is the soil-cement brick. Soil is the material used in the greatest proportion in the materials which compose this brick. Cement is a stabilizing material used to a lesser extent in the literature and has a content ranging from 4 to 30% by weight [7, 8]. However, Brazilian Portland Cement Association—ABCP [9] recommends the content of 5% as the lowest cement content to be adopted, while Sherwood [10] recommends a minimum content of 6%, since the soil is only stabilized at these levels and the properties are significantly improved.

Producing soil-cement bricks is performed with a manual or hydraulic press. The fresh soil-cement mixture is placed into molds and is pressed into the final product. The material is considered ecological and has gained visibility from an environmental point of view because, unlike traditional ceramic bricks, it does not require burning during the manufacturing process, which reduces the emission of greenhouse gases into the atmosphere. These have excellent thermal and acoustic properties and are considered suitable and readily available materials for building affordable houses [4]. Besides, soil-cement bricks have shown enormous potential for incorporating waste from various industries.

Research which uses solid waste in materials has the greatest appeal in the correct disposal of environmental liabilities which were previously dumped in inappropriate places causing pollution in the environment. A fact which often goes unnoticed is that some residues are able to improve the technical properties of new products in which they are incorporated into due to their intrinsic characteristics. In the case of soil-cement bricks, some wastes improve the properties of compression strength, water absorption and durability, as they cause changes in their macro- and microstructures.

The possibility of incorporating wastes in soil-cement bricks is shown by several works which used different wastes and achieved satisfactory results. As example, da Silva et al. [11] incorporated ceramic waste in soil-cement bricks; Ashour et al. [12] used natural fibers from wheat straw and barley straw in the formulations; de Ferreira and Cunha [13] partially replaced cement with rice husk residue in soil-cement bricks; Leonel et al. [14] investigated the possibility of inserting foundry sand into brick compositions; Medjo Eko et al. [15] inserted tire steel fibers; Paschoalin Filho et al. [16] studied soil-cement bricks with the incorporation of PET bottle waste; Campos et al. [17] manufactured soil-cement bricks using tertiary water from a sewage treatment plant; Barbosa et al. [18], incorporated rice husk and sludge from a water treatment plant; Islam et al. [19] used fly ash in conjunction with cement in formulations; and Reis et al. [20] analyzed the physical and mechanical properties of soil-cement bricks with the incorporation of quartzite mining waste.

The results obtained in studies of incorporating residues into soil-cement bricks demonstrate that the material is a viable solution for the correct destination of environmental liabilities which do not harm the technical properties of the new product and in some cases even improve the properties. Thus, soil-cement bricks are presented as a possible recipient for incorporating cassava wastewater during the manufacturing process, in which potable water can be replaced by the effluent.

It is important to remember that cassava wastewater is a source of pollution when disposed of inappropriately. This is evidenced by the enormous organic load found in the tests of chemical oxygen demand (COD), 14.7–101.38 gO_2/L, and biochemical oxygen demand (BOD), 6.21–29.20 gO_2/L, verified in the literature [21, 22]. Are generated about 300–600 L of cassava wastewater in flour houses, for every ton of roots processed [23, 24]. This shows the need for alternatives to treat or reuse the liquid.

Research in several areas has been carried out with cassava wastewater in order to reuse the effluent. Use in the animal diet [25]; fertilizer [26]; bio-hydrogen production [27]; insecticide and pesticide [28, 29], among others.

Moreover, cassava wastewater was also applied in construction materials. Araújo et al. [30] evaluated raw earth bricks with the effluent. Akpokodje and Uguru [31] produced blocks of sand with cement and liquid waste. However, research on cassava wastewater in building materials is rare. Therefore, investigations with the use of effluent in the production of these materials are necessary.

In this chapter are presented the results of a study in which were manufactured soil-cement bricks using the cassava wastewater from flour house. The main objective is to certify the technical feasibility of using cassava wastewater to replace potable water in formulations of solid soil-cement bricks with no specific structural function. Thus, will be addressed the following points to achieve this objective: an evaluation of the physical and mechanical behavior of the produced bricks, considering their water absorption and compression resistance properties; a study on the durability of the produced bricks; an evaluation of the microstructure of the best compositions using X-ray diffraction (XRD) and scanning electron microscopy (SEM) tests; and finally, a comparison of the results obtained by the soil-cement bricks manufactured with cassava wastewater with those produced with potable water.

2 Used Materials

The raw materials used in the research were soil, cement, water, and cassava wastewater. These are found easily in the state of Rio Grande do Norte (RN), Brazil.

2.1 Soil

The soil used to prepare the bricks was collected on the BR-101 (126 km) highway on the route between the cities of São José de Mipibú and Goianinha, near the Baldun River, in the State of Rio Grande do Norte, Brazil. It is a sandy-clayey soil (SC) according to the Unified Soil Classification System (USCS) of the American Society for Testing and Materials ASTM D2487 [32], characterized as A-4, with silt-clay fraction according to ASTM D3282 [33]. The material was chosen because it meets the geotechnical characteristics necessary for manufacturing soil-cement bricks as evaluated in granulometric composition tests and the Atterberg limits.

2.2 *Cement*

Was used the cement CP II F-40 (Portland cement composed with filler), which is a class of cement that contains a high limestone filler content. The material also has properties recommended by ABCP—Brazilian Portland Cement Association [9] for the production of soil-cement compositions.

2.3 *Water*

Water was used from the public supply system in the city of Natal, RN, Brazil, supplied by the local concessionaire CAERN (Water and Sewage Company of Rio Grande do Norte). It is potable water, with its potability parameters measured by the supply company itself.

2.4 *Cassava Wastewater*

The effluent was collected from small cassava flour houses located in the city of Lagoa Nova, RN, Brazil. The liquid was exposed to the sun to decrease the cyanide content present in the cassava wastewater composition, as mentioned in the introduction of this book.

3 Methodology

Four formulations for soil-cement bricks were analyzed in this study. The compositions were defined taking into account the minimum content of 6% of cement for soil stabilization established by Sherwood [10], and the significant results obtained by da Silva et al. [11] when using 12% cement to stabilize a similar soil. In addition, another criterion for defining the mixtures was to compare samples with total replacement of potable water by cassava wastewater with standard samples which only use water in the production process. The nomenclatures as well as the number of bricks that were manufactured for each formulation are shown in Table 1.

Table 1 Different analyzed compositions

Composition		Compressive strength (NBR 8492)		Water absorption (NBR 8492)		Durability (NBR 13554)	
		No of bricks	Age (days)	No of bricks	Age (days)	No of bricks	Age (days)
1	6C + 100CW	6	7 e 28	3	7	3	7
2	6C + 0CW	6	7 e 28	3	7	3	7
3	12C + 100CW	6	7 e 28	3	7	3	7
4	12C + 0CW	6	7 e 28	3	7	3	7
Total		24		12		12	
		48					

Note 6C: 6% of cement; 12C: 12% of cement; 100CW: 100% of cassava wastewater instead of potable water; 0CW: 0% of cassava wastewater instead of potable water

3.1 Materials Characterization

In the soil were conducted tests according to Brazilian Standards (NBR's); granulometric analysis, NBR 7181 [34]; compaction, NBR 7182 [35]; Atterberg limits, NBR 6459 [36] and NBR 7180 [37]. Furthermore, was performed the X-ray diffraction test (XRD) to identify the mineralogical phases of the soil, on the Shimadzu XRD-6000 device, with Cu-Kα radiation and scanning angle (2θ) from 10° to 70°.

The soil, cement, and cassava wastewater were subjected to X-ray fluorescence (XRF) assay, using the Shimadzu EDX-700 apparatus to determine the chemical composition. The settleable solids from the cassava wastewater were oven-dried at 60 °C for 24 h for use in the test.

The following characterization tests were performed in cassava wastewater: cyanide content, hydrogen potential (pH), suspended solids, total solids, macronutrients, and micronutrients. The cyanide content was determined at the Bioagri laboratory in the city of Paulista, Pernambuco (PE), Brazil, using as reference the standard of the American Society for Testing and Materials—ASTM D 7237 [38]; Hydrogen potential (pH), suspended solids and total solids at the Federal Institute of Education, Science and Technology of Paraíba (*IFPB*), in the city of João Pessoa, Paraíba (PB), Brazil and Tests of macronutrients and micronutrients in the laboratory of the Agricultural Research Company do Rio Grande do Norte (*EMPARN*), Parnamirim, RN, Brazil, being defined all parameters according to the Standard Methods for the Examination of Water and Wastewater of [39].

3.2 Molding and Curing of the Bricks

The solid soil-cement bricks were molded according to NBR 10833 [40]. A total of 48 bricks were manufactured, 12 for each formulation. The mixtures were mixed using a Motomil-MB-150L concrete mixer, and solid bricks were produced in a manual press (Sahara brand) with the following internal dimensions: 5 cm thick, 10 cm wide and 21 cm length, which configures the brick as type A between the two types (A and B) presented by NBR 8491 [41]. After pressing, the bricks went through a wet curing process using a sprinkler for 7 days in a covered location.

3.3 Simple Compressive Strength Test

Simple compressive strength tests of solid bricks were performed at 7 and 28 days. The samples were ruptured in an AMSLER universal testing machine with a load cell of 10,000 kg, according to NBR 8492 [42]. To perform the test, the bricks were first cut in half, and then joined by a cementitious paste and capped with the same paste, then after the paste dried, they were submerged in water for 6 h before the test was conducted.

3.4 Water Absorption Test

The water absorption tests were performed at 7 days after curing the samples. In this test, the bricks were weighed and placed in a FANEM greenhouse until they reached constant mass. Soon after, the bricks were weighed again and immersed in water for a period of 24 consecutive hours, then removed from immersion and weighed. The data obtained in the weighing, drying and water immersion cycles were used to calculate the absorption percentage of the pressed solid bricks. The tests followed the requirements of NBR 8492 [42].

3.5 Durability Test by Immersion and Drying

The durability tests by immersion and drying began at 7 days of curing the samples. The recommendations of NBR 13554 [43] were followed in conducting this test, however with some adaptations. The method prescribes six cycles with 5 h of wetting and 42 h of drying in an oven at a temperature of 71 ± 2 °C. The brushing step of the samples was not performed. Grande [44] reports that this step is not necessary in the bricks used in civil construction works due to requests for superficial abrasion to which they are submitted, being less rigorous than those found in pavement used for road purposes. This test ran for 12 days.

3.6 Mineralogical Analysis of the Compositions

The samples which showed more expressive results in the compressive strength tests were subjected to X-ray diffraction tests on a Shimadzu XRD-6000 apparatus with Cu-Kα radiation and scanning angle (2θ) from 10° to 70° to identify the mineralogical phases. The samples selected to perform these tests were immersed in absolute ethyl alcohol for 24 h and dried in an oven at 110 °C for a period of 6 h before the tests were performed, following the method of preparing samples for porosimetry by mercury intrusion [45]. Next, they were packed in hermetically sealed containers. The purpose of this procedure is to remove the free water present in the samples of the soil-agglomerate mixtures, inducing paralysis of hydration reactions (cement and pozzolanic), and delaying changes in the material's microstructure.

3.7 Scanning Electron Microscopy (SEM)

Morphological characterization of the compositions which showed the most significant results in the compressive strength tests was also performed to analyze the fracture surface of the samples. The images were obtained by scanning electron microscopy (SEM) with an acceleration voltage (EHT) of 3 kV, working distance (WD) from 3.1 to 3.2 mm, and image magnification from 20 to 30 KX.

4 Results and Discussions of the Characterization of Materials

In this topic are presented the results of the characterization of the materials. These serve as a basis for understanding the behavior of bricks in physical, mechanical, and microstructural tests.

4.1 Soil Characterization

Figure 1 shows the granulometric curve of the soil used. The soil has a passing percentage of 100% in a No. 4 sieve (4.8 mm), 69.55% in a No. 40 sieve (0.42 mm), and 36% in a No. 200 sieve (0.075 mm).

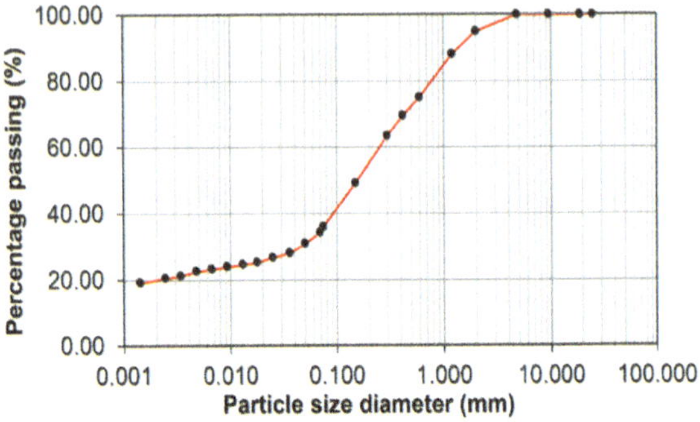

Fig. 1 Grain size distribution curve of the soil

Table 2 Atterberg limits and specific mass of soil

Parameter	Water (%)	Cassava wastewater (%)
Liquid limit (LL)	25.30	22
Plasticity limit (PL)	18.10	14.30
Plasticity index (PI)	7.20	7.70

The curve shows that the soil used meets the requirements of NBR 10833 [40] for soils used in soil-cement compositions, which are: 100% of the particles passing through a 4.8-mm-mesh sieve; and 10–50% passing through a 0.075-mm-mesh sieve opening. The percentage of 36% passing through a No. 200 sieve (0.075 mm) shows a significant amount of fine matter in the soil, which indicates the presence of clay minerals. These clay minerals have a reduced diameter and can provide a filler effect responsible for refining the pores, leaving the material denser with reduced water absorption and significant resistance to compression.

The results of the Atterberg limits in Table 2 show that the values met the NBR 10833 standard [40]. The standard establishes that soils used in soil-cement compositions must have a liquidity limit equal to or less than 45% and a plasticity index equal to or less than 18%.

The compaction test was performed according to the NBR 7182 standard [35], to determine the water content to be used in preparing the mixtures. The results are displayed by means of the curves plotted in Fig. 2.

Compactions carried out with cassava wastewater and water showed 10.01 and 13.53% of optimum humidity and 1.99 and 1.90 g/cm^3 of dry bulk density, respectively. The soil composition with cassava wastewater has a lower optimum moisture content associated with a higher dry bulk density, which indicates better packaging of the mixture particles.

Finally, chemical composition tests were performed on the soil by X-ray fluorescence (XRF), and a mineralogical analysis was performed by an X-ray diffraction (XRD) assay. The XRF results are shown in Table 3.

In the values obtained for the soil under study, it is possible to observe high levels of silica (SiO_2) and alumina (Al_2O_3), typical of lateritic soils. When in a disorganized

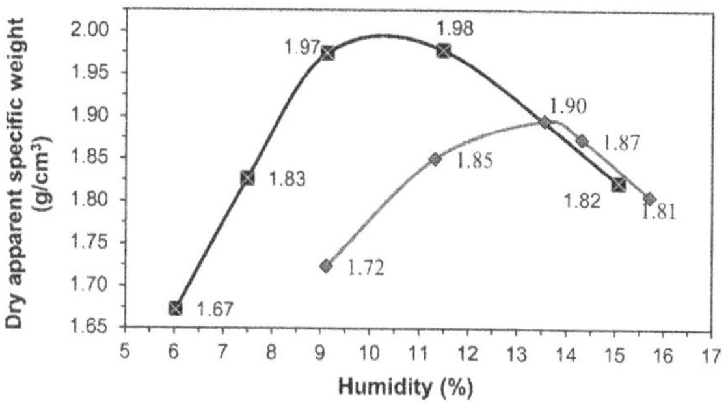

Fig. 2 Soil compaction curves with cassava wastewater and water

Table 3 Chemical composition of the soil

Oxide	SiO_2	Al_2O_3	Fe_2O_3	K_2O	CaO	TiO_2	ZrO_2	BaO
Weight (%)	40.494	32.611	19.804	0.140	0.329	1.853	4.104	0.460

form, the silica and alumina minerals present in the soil may recombine with the portlandite (Ca $(OH)_2$) formed during the cement hydration process and produce calcium silicate hydrate (CSH) and/or calcium alumina silicate hydrate (CASH), which are responsible for the mechanical strength gain in materials which use cement in their formulation. The results of the XRD assay were plotted on a graph and shown in Fig. 3.

The kaolinite and quartz mineralogical phases were identified in accordance with the high silica and alumina levels previously presented in Table 3. The presence of the kaolinite and quartz minerals is justified by the fact that it is a lateritic soil, typical of tropical regions.

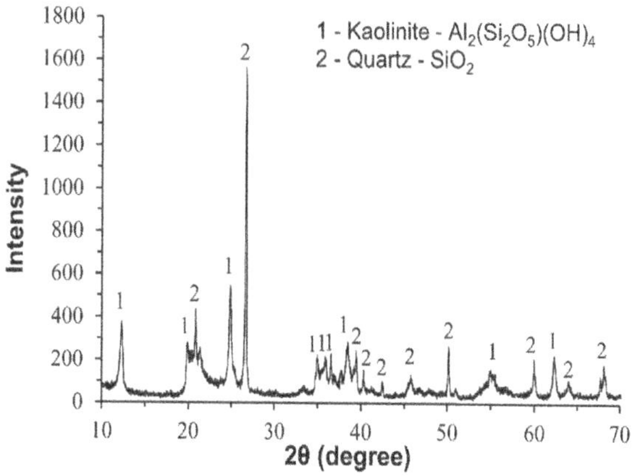

Fig. 3 X-ray diffraction pattern of the soil

Table 4 Chemical composition of the cement

Oxide	SiO_2	Al_2O_3	Fe_2O_3	CaO	MgO	SO_3
Weight (%)	12.796	2.738	6.727	73.275	1.771	2.694

4.2 Cement Characterization

Table 4 shows the chemical composition of the cement used obtained through the XRF assay.

According to the results for the CP II-F40 cement used, there is a material with high calcium (CaO) and silica (SiO_2) content in its composition which is characteristic of cements with a high concentration of limestone filer.

4.3 Cassava Wastewater Characterization

Table 5 shows the chemical composition of the cassava wastewater used in the research with soil-cement bricks.

The verified pH shows the acidity of the effluent used. The liquid also has similarities with the work of Silva et al. [22], with high concentrations of potassium, nitrogen, sodium, iron, and calcium. In this study were not performed the COD

Table 5 Chemical composition of cassava wastewater

Parameters analyzed	Values
Nitrogen (mg/L)	1121
Phosphorus (mg/L)	132
Potassium (mg/L)	1456
Calcium (mg/L)	93
Magnesium (mg/L)	219
Sodium (mg/L)	351
Zinc (mg/L)	20
Copper (mg/L)	9
Iron (mg/L)	117
Manganese (mg/L)	15
Total solids (g/L)	14.56
Suspended solids (g/L)	2.37
Chloride (mg/L)	545.4
Sulfur (mg/L)	78.53
Phosphate (mg/L)	27.07
Cyanide (mg/L)	5.62
pH	4.5

Table 6 Chemical composition of the powder from cassava wastewater

Oxide	SiO_2	Al_2O_3	Fe_2O_3	K_2O	CaO	SO_3	P_2O_5	Na_2O	Br
Weight (%)	2.457	2.306	15.922	32.654	25.739	3.123	3.592	9.986	4.220

and BOD tests, as the scope of this book does not include the survey of parameters to assess the biodegradability of this liquid waste and the wastewater treatment methods to be used, but the application of the environmental liability for the manufacture of alternative products for the construction industry. However, the results of COD and BOD presented in the literature demonstrate the high organic load of cassava wastewater [21, 22]. Table 6 shows the results obtained in the XRF test.

The chemical composition obtained in the XRF test demonstrates that potassium (K_2O), calcium (CaO), iron (Fe_2O_3), and sodium (Na_2O) make up most of the powder, and these can increase the number of exchangeable ions in the mixture of soil-cement and cassava wastewater.

5 Results and Discussions of Technological Tests

The simple compressive strength test was conducted to check the mechanical strength of the bricks when applying compressive loads. Meanwhile, water absorption and durability tests served to verify the physical behavior of the bricks with regard to the porosity for the absorption test and the ability to resist the weather in the durability test.

5.1 *Simple Compressive Strength*

Figure 4 presents the results obtained for the compositions analyzed in the simple compressive strength test. It can be seen that all values were above the recommendation of the NBR 13553 [46], which establishes a minimum of 1 MPa for materials used in the construction of soil-cement walls without structural purposes.

Compositions with 12% of cement showed greater compression strength. It can be explained by the fact that a high binder content is responsible for a more significant formation of hydrated cementitious compounds, which are directly linked to the gain in mechanical strength. According to Islam et al. [19], with the increase in the amount of cement, more CSH gels are produced, which unite the soil particles, thus increasing the strength. In comparing the samples produced with cassava wastewater to those manufactured with potable water in the same binder proportion (6 and 12%), those that had cassava wastewater obtained higher average results at all analyzed ages. However, when considering the standard deviation, it is possible to notice that the results presented by both samples are similar.

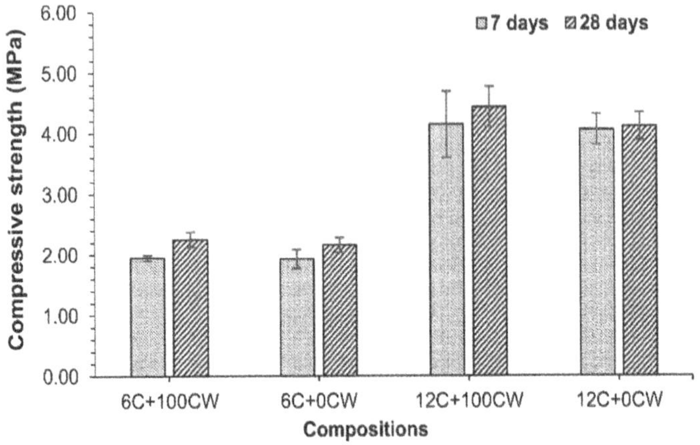

Fig. 4 Compressive strength of the analyzed samples

Despite the low pH (4.5) presented by cassava wastewater which is considered acidic, in a work carried out in the agronomic area by Barreto et al. [47], the authors state that the pH increases due to the high percentage of cations when the cassava wastewater is added to the soil, which also provides an increase in the cation exchange capacity of the medium. In the same study, the authors found that cassava wastewater also provided an increase in exchangeable calcium. Similar phenomena are observed in correlating this with civil engineering when using lime to stabilize soils; according to Vizcarra [48], the pH and the amount of free calcium cations increase when using lime in the soil. Calcium cations can replace other cations within the cation exchange complexes that occur in the soil. The author also states that the cation exchange process is partially responsible for agglomerating and flocculating soil particles which are treated with lime, and the same phenomenon was observed in this study when preparing the mixtures, evidencing the binding characteristic of the cassava wastewater.

The change in pH and the possible increase in the capacity for cation exchange and free calcium cations may be responsible for the small increase in the average value of the compressive strengths presented by the samples containing the cassava wastewater compared to those containing water in the composition. The influence of cassava wastewater during soil pressing is another factor that may have corroborated to the small average increase in compressive strength. In the soil compaction test, presented previously in Sect. 4.1, when was used the effluent to replace water, the sample under study showed a higher apparent density. The liquid waste allows a more efficient pressing with higher accommodation of the particles of the cement soil. Consequently, more compact structures tend to be more resistant.

It is worth mentioning that cassava wastewater did not affect the cement hydration process. Therefore, the substitution of water for the effluent in manufacturing the bricks is completely feasible, thus giving cassava wastewater an environmentally correct destination, and at the same time enables potable water to be used for more noble purposes such as human and agricultural consumption.

5.2 Water Absorption

The results of the water absorption test are shown in Fig. 5. It is possible to observe that all studied compositions presented water absorption values below 20%, which is established as a maximum limit for soil-cement formulations by the NBR 8491 standard [41].

In a study which also used agro-industrial residues in soil-cement formulations, de Ferreira and Cunha [13] also obtained water absorption values below 20%. On the other hand, not all formulations met the 20% limit in a study carried out by Siqueira and Holanda [49] with agro-industrial residue. The compositions with 12% cement presented the lowest water absorption levels due to the higher binder percentage providing more cement reactions, resulting in denser compositions with reduced pore size. Mengue et al. [50] state that CSH is the main responsible for the cohesion and stabilization properties of soil-cement material.

The influence of cassava wastewater during soil-cement compaction can also justify the lower average absorption values presented by the mixtures with the residue. The more compact structures have fewer voids, which contributes to reducing water absorption.

The results presented can be correlated with the compressive strength test, since the higher the uniaxial strengths, the lower the average water absorption levels due to the decrease in the porosity of the evaluated mixtures. Despite showing a lower

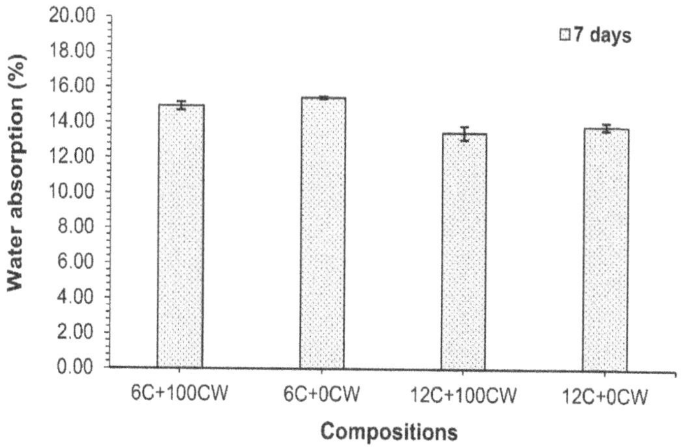

Fig. 5 Water absorption of the analyzed samples

average water absorption when comparing the formulations with cassava wastewater, the standard deviation results were similar, which indicates that the use of the wastewater did not harm this physical property.

5.3 Modified Durability

The mass loss results after the bricks were submitted to the modified durability test shown in Fig. 6, showing that the mass loss values were well below the maximum limit of 10% established by NBR 13553 [46] for A-4 soils.

Efficient pressing of the soil together with the chemical reactions provided by applying the stabilizing agent (cement) can explain the reduced mass loss values because mechanical stabilization is provided by the compaction in the first case, while the cement reactions give chemical stabilization to the soil in the second case. These facts were also found in the study by da Silva et al. [11], in which the authors stabilized a similar soil by pressing and using cement as a stabilizing agent, and obtaining low mass loss values in the analyzed samples.

The results obtained in this study are inferior to those obtained in the study by Siqueira et al. [51], in which industrial solid waste was used in soil-cement formulations. França et al. [52] also obtained results of mass loss below 10% in research with the incorporation of limestone residue from marble processing in soil-cement bricks.

It is important to emphasize that the bricks with cassava wastewater showed less mass loss when compared to those which contained water in the formulation, with these samples probably having a denser structure. The greater cohesion between the soil-cement particles provided by cassava wastewater during pressing makes it challenging to detach parts of the material. In this way, the bricks are more resistant

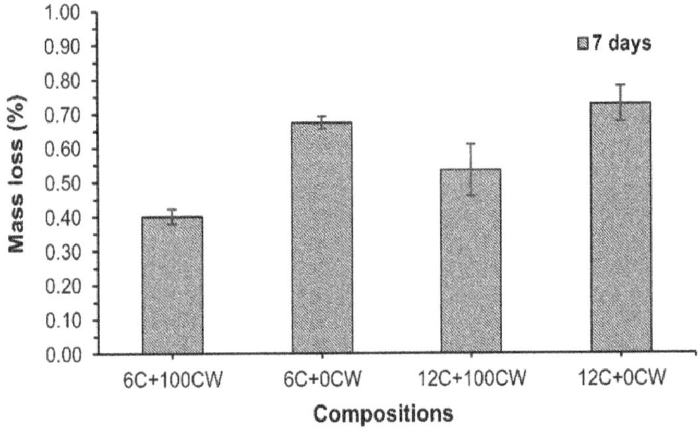

Fig. 6 Mass loss of analyzed samples

to the weather to which they will be subjected during their useful life. So, the bricks with the effluent are more durable than those produced with water.

6 Results and Discussions of Microstructural Tests

X-ray diffraction (XRD) and scanning electron microscopy (SEM) tests were performed on the bricks, showing the best results in the simple compression resistance test. These tests were intended to evaluate the microstructure of the bricks through mineralogical and morphological analysis, respectively.

6.1 Mineralogical Analysis by XRD

The selected samples were those with 12% cement at 28 days with and without cassava wastewater. Figure 7 shows the X-ray diffraction of the composition with 12% cement and 100% cassava wastewater to replace water. It is possible to verify the predominance of kaolinite clay and quartz in the soil XRD. The calcite phase found can be justified by the high calcium percentage presented in the cement composition, which undergoes a carbonation process when in contact with the air. CSH and CASH were also found, which are phases resulting from the cement hydration process. It is known that the observed cement phases contribute to the increase in mechanical strength. CSH was found in the composition of $Ca_5(Si_6O_{16})(OH)_2$, known as clinotobermorite, which was also found by Braz et al. [53] when they used aluminum recycling residues, sugarcane bagasse ash and zeolite in mortars. CASH was found with the composition of $Ca(Al_2Si_2O_7(OH)_2)(H_2O)$, called lawsonite, also found by

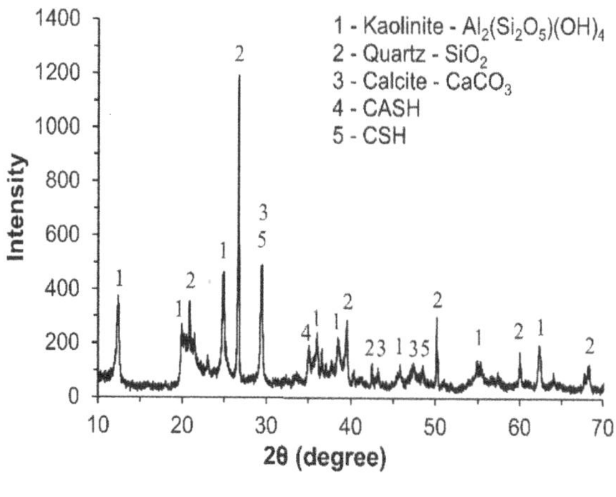

Fig. 7 X-ray diffraction pattern of composition 12C + 100CW at 28 days

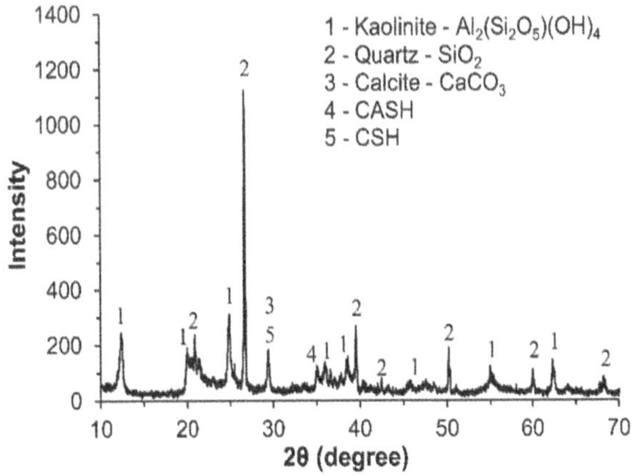

Fig. 8 X-ray diffraction pattern of composition 12C + 0CW at 28 days

Kupwade-Patil et al. [54] when studying the microstructure of cement pastes using pozzolanic volcanic ash.

The X-ray diffraction of the sample with 12% cement without cassava wastewater is shown in Fig. 8. The same phases were observed in the mixture with cassava wastewater, which shows that the use of the effluent did not impair cement reactions. However, the CSH peak at position 48.55°2θ observed in the sample with cassava wastewater does not have a significant intensity in this mixture. Kaolinite and quartz still predominate in diffraction due to the fact that the soil is used in a greater proportion in formulations. Another relevant fact is that the peaks appear to be less intense, and consequently, fewer peaks were identified in the formulation with water when compared to the formulation which used cassava wastewater, which may indicate that the sample with cassava wastewater has significant crystallinity.

It can also be seen that the CSH phase appears together with the calcite phase at approximately 20.46°2θ in both formulations. This is due to an overlap of peaks in this region.

6.2 *Scanning Electron Microscopy (SEM)*

As in the X-ray diffraction test, the compositions which showed the best results in the compressive strength tests were subjected to scanning electron microscopy (SEM). Samples with the same ages and nomenclatures from the X-ray diffraction test were used, 12C + 100CW and 12C + 0CW.

The SEM of the formulation with 12% cement and cassava wastewater is shown in Fig. 9. Some spongy formations characteristic of the CSH can be seen in addition to typical quartz particles of the lateritic soil used. Similar formations were presented in the work of Braz et al. [53] for CSH. Some fibrous structures were confirmed in

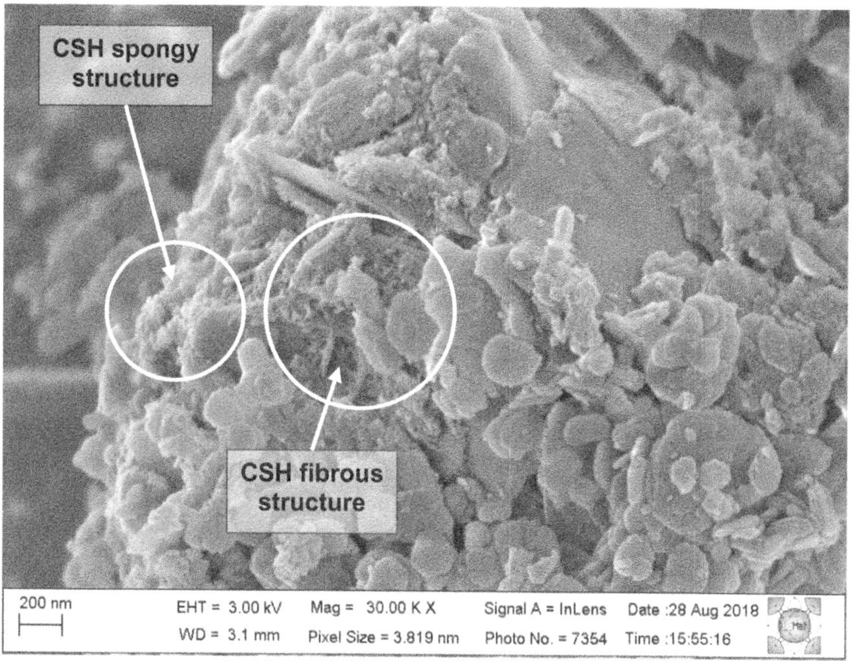

Fig. 9 SEM of composition 12C + 100CW at 28 days magnification 30,000×

the formulations. According to Da Fonsêca [55], these structures may be related to CSH and contribute to the gain in the sample's mechanical strength.

The micrograph in Fig. 10 shows the microstructure of the sample in which cassava wastewater was not used to replace water. The same spongy and fibrous structures characteristic of CSH are observed, which shows that the cassava wastewater did not impair the cement hydration process. This is in accordance with the previously presented physical and mechanical tests. Islam et al. [19] also verified the presence of CSH gels in the study with the application of fly ash in conjunction with cement to stabilize earth blocks.

Both samples generally showed structures which indicate the presence of CSH, the main product of cementitious reaction responsible for increasing the mechanical strength in cement-based materials. The CSH also fills empty spaces at the same time that it binds soil particles, contributing to the low values of water absorption and mass loss, which justifies the previously observed results.

7 Conclusions

This chapter had the main objective to certify the technical viability of substituting potable water with cassava wastewater in formulations of solid soil-cement bricks

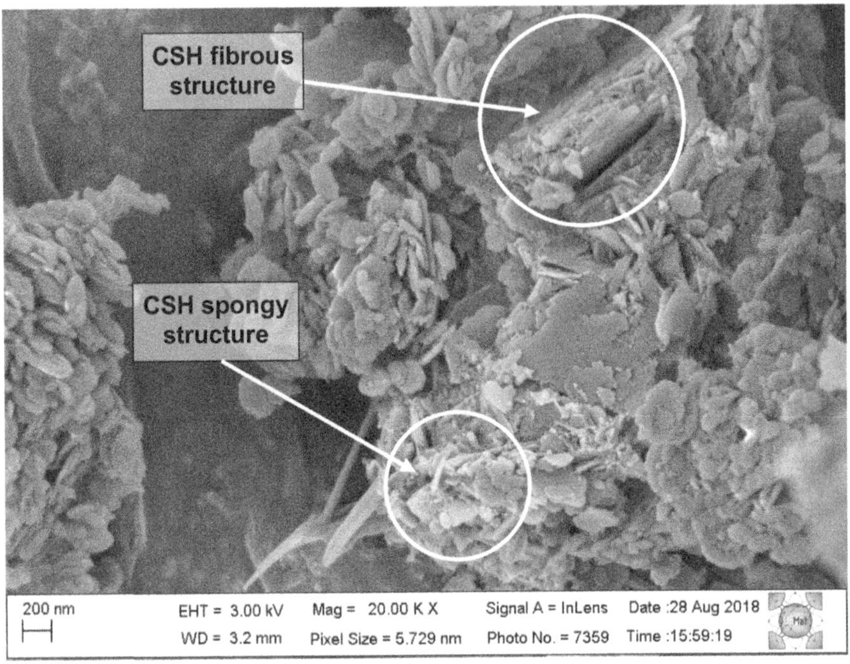

Fig. 10 SEM of composition 12C + 0CW at 28 days magnification 20,000×

without specific structural function. The results obtained in the technical tests showed that the bricks analyzed met the requirements of Brazilian standards for compression strength, water absorption, and loss of mass.

When comparing the results obtained by the soil-cement bricks with cassava wastewater to those which used water in the mixture, the bricks with cassava wastewater showed similar results in the compressive strength and water absorption tests and lower mass losses in the durability test, which proves the technical viability of producing soil-cement bricks with cassava wastewater and their use in masonry without structural purposes.

From a technical point of view, substituting water for cassava wastewater was shown to be viable for producing soil-cement bricks, since cassava wastewater does not harm the hydration process of cement and the consequent formation of cementitious compounds over time. From an environmental point of view, there is the great advantage of not using potable water for manufacturing bricks, as it is a scarce commodity in many regions of the Brazilian semi-arid region.

The development of the new material presented in this chapter resulted in an invention patent application at the National Institute of Intellectual Property (INPI) registered under number BR 10 2019 003002 0.

The results presented throughout this chapter attest to the feasibility of applying the environmental liability (cassava wastewater) to soil-cement bricks; however, the following considerations must be observed: a previous treatment must be conducted on the cassava wastewater to reduce the cyanide content by exposing the effluent to the sun, thus suggesting that open tanks should be built in large-scale productions to store the waste; over time the cassava wastewater begins to present an increasingly intense odor when it is not refrigerated, and this intensifies the action of microorganisms, therefore requiring the use of masks to handle the liquid. Thus, it is recommended that the effluent be stored in refrigerators to prolong its useful life. Although these soil-cement bricks are considered ecological, extracting soil for larger productions deserves attention, as large earth movements modify the landscape and degrade the environment, and therefore materials which allow partial or total replacement of the soil must be idealized.

Acknowledgements This study was financed in part by the Coordenacao de Aperfeicoamento de Pessoal de Nivel Superior—Brasil (CAPES)—Finance code 001, the Federal Institute of Education, Science and Technology of Paraiba and the Federal University of Rio Grande do Norte.

References

1. Seco A, Urmeneta P, Prieto E et al (2017) Estimated and real durability of unfired clay bricks: determining factors and representativeness of the laboratory tests. Constr Build Mater 131:600–605
2. Bernat-Maso E, Gil L, Escrig C (2016) Textile-reinforced rammed earth: experimental characterisation of flexural strength and thoughness. Constr Build Mater 106:470–479
3. Jannat N, Hussien A, Abdullah B, Cotgrave A (2020) Application of agro and non-agro waste materials for unfired earth blocks construction: a review. Constr Build Mater 254:119346
4. Mimboe AG, Abo MT, Djobo JNY et al (2020) Lateritic soil based-compressed earth bricks stabilized with phosphate binder. J Build Eng 31:101465
5. Ouedraogo KAJ, Aubert J-E, Tribout C, Escadeillas G (2020) Is stabilization of earth bricks using low cement or lime contents relevant? Constr Build Mater 236:117578
6. Pavan GS, Ullas SN, Nanjunda Rao KS (2020) Interfacial behavior of cement stabilized rammed earth: experimental and numerical study. Constr Build Mater 257:119327
7. da Silva Segantini AA, Wada PH (2011) An evaluation of the composition of soil cement bricks with construction and demolition waste. Acta Sci Technol 33:179–183
8. Vinai R, Lawane A, Minane JR, Amadou A (2013) Coal combustion residues valorisation: research and development on compressed brick production. Constr Build Mater 40:1088–1096
9. Brazilian Portland Cement Association—ABCP (2004) Technical Study 35 - Dosage of Soil-Cement Mixtures: Dosing Standards and Test Methods, São Paulo
10. Sherwood PT (1993) Soil stabilization with cement and lime: state-of the-art review. Transp Res Lab Her Majesty's Station Off London
11. da Silva VM, Góis LC, Duarte JB et al (2014) Incorporation of ceramic waste into binary and ternary soil-cement formulations for the production of solid bricks. Mater Res 17:326–331
12. Ashour T, Korjenic A, Korjenic S, Wu W (2015) Thermal conductivity of unfired earth bricks reinforced by agricultural wastes with cement and gypsum. Energy Build 104:139–146
13. de Ferreira RC, Cunha AHN (2017) Quality evaluation of soil-cement-plant residue bricks by the combination of destructive and non-destructive tests. Rev Bras Eng Agrícola e Ambient 21:543–549
14. Leonel RF, Folgueras MV, Dalla Valentina LVO et al (2017) Characterization of soil-cement bricks with incorporation of used foundry sand. Cerâmica 63:329–335

15. Medjo Eko R, Offa ED, Yatchoupou Ngatcha T, Seba Minsili L (2012) Potential of salvaged steel fibers for reinforcement of unfired earth blocks. Constr Build Mater 35:340–346
16. Paschoalin Filho JA, Storopoli JH, Guerner Dias AJ (2016) Evaluation of compressive strength and water absorption of soilcement bricks manufactured with addition of pet (polyethylene terephthalate) wastes. Acta Sci Technol 38:163–171
17. Campos A, Nascimento Junior JB do, Brito LT (2019) Structural behavior of soil-cement bricks using different sources of water and healing methods. Interações Campo Gd, vol 20, pp 283–296
18. Barbosa MFL, Pironcelli ABS, Silva CA et al (2019) Rice husk and water treatment plant sludge incorporated into soil–cement brick. Asian J Civ Eng 20:563–570
19. Islam MS, Elahi TE, Shahriar AR, Mumtaz N (2020) Effectiveness of fly ash and cement for compressed stabilized earth block construction. Constr Build Mater 255:119392
20. Reis FMD, Ribeiro RP, Reis MJ (2020) Physical-mechanical properties of soil-cement bricks with the addition of the fine fraction from the quartzite mining tailings (State of Minas Gerais, Brazil). Bull Eng Geol Environ 79:3741–3750
21. Peres S, Monteiro MR, Ferreira ML et al (2019) Anaerobic digestion process for the production of biogas from cassava and sewage treatment plant sludge in Brazil. BioEnergy Res 12:150–157
22. Silva FF, Bertonha A, Freitas PSL et al (2005) Fragment of starch to glucose to identify waste water from manioc industry in depth of soil. Acta Sci Biol Sci 27:507–511
23. Carvalho JC, Borghetti IA, Cartas LC et al (2018) Biorefinery integration of microalgae production into cassava processing industry: Potential and perspectives. Bioresour Technol 247:1165–1172
24. Ribas MMF, Cereda MP, Villas Bôas RL (2010) Use of cassava wastewater treated anaerobically with alkaline agents as fertilizer for maize (Zea mays L.). Brazilian Arch Biol Technol 53:55–62
25. Santana Neto JA, Martinele I, Cedrola F et al (2017) Apparent digestibility and rumen protozoal profile of sheep fed cassava wastewater. Rev Bras Saúde e Produção Anim 18:327–336
26. de Araújo NC, de Lima VLA, de Lima GS et al (2019) Nutrient contents and growth of corn fertigated with human urine and cassava wastewater. Rev Bras Eng Agrícola e Ambient 23:681–686
27. Meier TRW, Cremonez PA, Maniglia TC et al (2020) Production of biohydrogen by an anaerobic digestion process using the residual glycerol from biodiesel production as additive to cassava wastewater. J Clean Prod 258:120833
28. de Carvalho JD, Neves FL, Silva CD da, Bittencourt MAL (2017) Biological aspects and insecticide action of plant species on eggs and nymphs of citrus black fly (Aleurocanthus woglumi Ashby - Aleyrodidae) at laboratory level. Rev Bras Frutic 39:e-045
29. Zevallos DMP, Pereira Querol M, Ambrogi BG (2018) Cassava wastewater as a natural pesticide: current knowledge and challenges for broader utilisation. Ann Appl Biol 173:191–201
30. Araújo NC, Ramos AP, Queiroz AJP, dos Santos RC (2015) Mechanical properties of manufactured bricks with soil and cassava wastewater. Rev Verde 10:213–217
31. Akpokodje IO, Uguru H (2019) Compressive strength of commercially produced sandcrete blocks within isoko metropolis of delta state, Nigeria. Turkish J Agric Eng Res 1:91–103
32. ASTM (2017) ASTM D2487. Standard practice for classification of soils for engineering purposes (Unified Soil Classification System). ASTM international, West Conshohocken
33. ASTM (2015a) ASTM D3282. Standard practice for classification of soils and soil-aggregate mixtures for highway construction purposes. ASTM International, West Conshohocken
34. Brazilian Association of Technical Standards—ABNT (2016a) NBR 7181, Soil - Granulometric analysis, Rio de Janeiro
35. Brazilian Association of Technical Standards—ABNT (2016b) NBR 7182, Soil - Compaction test. Rio de Janeiro
36. Brazilian Association of Technical Standards—ABNT (2016c) NBR 6459, Soil - Determination of the liquidity limit, Rio de Janeiro
37. Brazilian Association of Technical Standards—ABNT (2016d) NBR 7180, Soil - Determination of the plasticity limit, Rio de Janeiro
38. ASTM (2015b) ASTM D7237. Standard test method for free cyanide and aquatic free cyanide with flow injection analysis (FIA) utilizing gas diffusion separation and amperometric detection. ASTM International, West Conshohocken

39. APHA (1998) Standard methods for the examination of water and wastewater, 20th edn. American Public Health Association, Washington
40. Brazilian Association of Technical Standards—ABNT (2012a) NBR 10833, manufacture of brick and block of soil-cement using manual or hydraulic press – Procedure, Rio de Janeiro
41. Brazilian Association of Technical Standards—ABNT (2012b) NBR 8491, Soil-cement brick – Requirements, Rio de Janeiro
42. Brazilian Association of Technical Standards—ABNT (2012c) NBR 8492, Soil-cement brick - Dimensional analysis, determination of compressive strength and water absorption - Test method, Rio de Janeiro
43. Brazilian Association of Technical Standards—ABNT (2012d) NBR 13554, Soil-cement - Wetting and drying durability test - Test method, Rio de Janeiro
44. Grande FM (2003) Production of soil-cement modular bricks by manual pressing with and without silica fume addition. Master dissertation. University of São Paulo. São Paulo/SP, 180 p
45. Duart MA (2008) Concrete microstructure study with addition of unground residual rice husk ash without burn control. Master dissertation. Universidade Federal de Santa Maria. Santa Maria/RS, 134 p
46. Brazilian Association of Technical Standards—ABNT (2012e) NBR 13553: Materials for use in monolithic soil-cement wall without structural function – Requirements, Rio de Janeiro
47. Barreto MTL, Rolim MM, Pedrosa EMR et al (2013) Chemical attributes of two soils under cassava wastewater application. Rev Bras Ciências Agrárias 8:528–534
48. Vizcarra GOC (2010) Applicability of municipal solid waste ash for pavements base. Master dissertation. Universidade Católica do Rio de Janeiro. Rio de Janeito/RJ, 120 p
49. Siqueira FB, Holanda JNF (2013) Reuse of grits waste for the production of soil–cement bricks. J Environ Manag 131:1–6
50. Mengue E, Mroueh H, Lancelot L, Medjo Eko R (2017) Physicochemical and consolidation properties of compacted lateritic soil treated with cement. Soils Found 57:60–79
51. Siqueira FB, Amaral MC, Bou-Issa RA, Holanda JNF (2016) Influence of industrial solid waste addition on properties of soil-cement bricks. Cerâmica 62:237–241
52. França BR, Azevedo ARG, Monteiro SN et al (2018) Durability of soil-cement blocks with the incorporation of limestone residues from the processing of marble. Mater Res 21:1–6
53. Braz IG, Shinzato MC, Montanheiro TJ et al (2018) Effect of the addition of aluminum recycling waste on the pozzolanic activity of sugarcane bagasse ash and zeolite. Waste and Biomass Valorization, pp 3493–3513
54. Kupwade-Patil K, Al-Aibani AF, Abdulsalam MF et al (2016) Microstructure of cement paste with natural pozzolanic volcanic ash and Portland cement at different stages of curing. Constr Build Mater 113:423–441
55. Da Fonsêca NJM (2018) Potential for replacing cement with lime in soil-cement bricks with incorporation of ceramic waste. Masters dissertation. Federal University of Rio Grande do Norte. Technology Center. Graduate Program in Civil Engineering. Natal/RN, 86 p

Ecological Brick of Mineral Residues and Cassava Wastewater

Ricardo Eugênio Barbosa Ramos Filho

Abstract This chapter deals with the total substitution of mixing water by cassava wastewater in manufacturing of ecological bricks, aiming at using environmental liabilities by using a combination of scheelite residue and rock dust, eliminating the use of soil which can also cause environmental impacts via its extraction. The cassava wastewater added plasticity and workability to the mixture, producing an ecological brick with similar characteristics to soil-cement bricks.

Keywords Ecological brick · Scheelite residue · Rock dust · Cassava wastewater

1 Introduction

As presented in the previous chapter, soil-cement bricks have numerous advantages over conventional bricks, as they do not need to be fired during the process, reducing the emission of greenhouse gases, reducing the use of mortar to bind the pieces, are easy to produce, and can be produced close to the work projects, also reducing the total cost. Despite these innumerable advantages, the production of soil-cement bricks can cause environmental impacts due to the removal of soil, which in its greatest amount can reach the range of 80–95% according to the Brazilian Portland Cement Association [1].

In addition to the environmental impacts caused by soil extraction, there is a difficulty in selecting and preparing it for use in civil construction, since it is a heterogeneous material. In order to standardize the use of soil in the manufacture of materials for civil construction, ABNT (Brazilian Association of Technical Standards) determined geotechnical parameters for the raw material to be applied in soil-cement.

R. E. B. Ramos Filho (✉)
Graduate Program in Materials Science and Engineering (PPGCEM), Technology Center (CT), Federal University of Rio, Grande do Norte (UFRN), University Campus, Natal, Rio Grande do Norte, Brazil
e-mail: ricramosf30@ufrn.edu.br

W. Acchar and V. Monteiro da Silva (eds.), *Use of Cassava Wastewater and Scheelite Residues in Ceramic Formulations*, Green Energy and Technology,
https://doi.org/10.1007/978-3-030-58782-6_3

These help from the selection of the soil to the preparation of the mixture and are related to granulometry, plasticity, consistency, among other characteristics.

An example of a standard for soil-cement is the NBR 13553 standard [2] on materials for use in monolithic soil-cement walls without structural function, in which the soil must present:

- 100% of material passing through the sieve with 4.75 mm mesh opening;
- Between 15 and 50% of material passing through the sieve with 75 μm mesh opening;
- Liquidity limit less than or equal to 45%;
- Plasticity index less than or equal to 18%.

A study of the combinations between scheelite residue, rock dust, and cassava wastewater was carried out with the addition of binders through correlations between microstructure, properties, and performance, with the justification of producing an innovative ecological brick promoting greater sustainability and economy through implementing residues and eliminating the use of soil.

2 Residues in Soil-Cement Bricks

The use of residues in soil-cement bricks has intensified in recent years due to satisfactory results, demonstrating that the product easily receives additives, and substitutes which do not harm the physical and mechanical properties, and in some cases, even improves them. There is an increase in the amount of research on residues from mineral and vegetal extraction, probably due to the high volume of waste resulting from this productive sector and the obtaining of promising results when its physical and mechanical properties are evaluated. Mineral residues that have low levels of clay minerals usually have reduced plasticity and consistency that can affect the production of soil-cement bricks.

The plasticity of clay minerals is the result of the lubricating action of water and the force of attraction between the lamellar particles of their crystalline structures [3]. Therefore, it can be assumed that plasticity develops when the clay has enough water to cover the entire accessible surface of the clay minerals, forming a film that facilitates the sliding of the plates of the crystalline structure.

It is possible to verify the use of residues in soil-cement bricks such as:

- Partial substitutes for binders, such as the use of rice husk ash in the work of de Ferreira and Cunha [4], construction waste researched by da Silva et al. [5], sugarcane bagasse ash in Faria et al. [6];
- Partial or total substitutes for hydration water as in the work of Araújo et al. [7] in his work on solid soil bricks concluded that the exchange of water for cassava wastewater in the ceramic mass does not negatively influence the mechanical properties;

- Partial substitute of the soil as in the work of [8] that used the residue of oil well drilling gravel from the drilling of onshore wells in total replacement of soil, and Paschoalin Filho et al. [9] that incorporated polyethylene terephthalate (PET), from PET bottles, into the soil, obtaining satisfactory results for soil-cement bricks without structural function.

3 Used Materials

The materials used in this research were residues in substitution to the soil (solid residues of the mineral extraction) and substitution to the water of hydration, by the use of the cassava wastewater, and also acting as an additive providing plasticity to the mixture.

3.1 Solid Residues

According to data from the National Department of Mineral Production (DNPM), scheelite production ($CaWO_4$) in Brazil corresponds to 0.5% of world production [10]. Despite the low worldwide production of scheelite, Brazil has the largest scheelite mine in Latin America, the Brejuí Mine, and the highest concentration of tungsten ore is found in the Scheelífera do Serido Province, in the states of Rio Grande do Norte and Paraíba (Brazil 2018).

Rock dust is the residue generated by the rock breaking process (crushing) to obtain crushed rock. As any rock with high mechanical resistance can be used as gravel in civil construction, there are several types of gravel with different geological formation. According to Bertolino et al. [11], due to the geological characteristics of the Brazilian territory, there is a great diversity of rocks used as aggregates, depending on local or regional availability such as granites, gneisses, basalt, and limestone. This residue has been widely used in research, as in the work of Machado et al. [12], with satisfactory results in the incorporation of ceramic, concrete, and clay mass, being used in the market as a substitute for natural sand.

3.2 Cassava Wastewater

Cassava wastewater is the effluent from the processing of cassava. It has a high degradation power due to the amount of organic matter and toxicity due to the high levels of cyanide and other acids. The toxicity of this effluent represents a major environmental problem when it is not treated before being discarded [13]. It is estimated that each ton of processed cassava is generated from 300 to 600 L of cassava wastewater [14].

This effluent has been used in several products such as fertilizer, pesticide, vinegar, alcohol, and bricks as in the work of Araújo [7] to replace hydration water and improve mechanical properties, obtaining an even more sustainable brick.

4 Methodology

The experimental procedure started with the collection and characterization of the materials, followed by homogenization, compacting, and molding to be carried out the technological tests of water absorption and resistance to simple compression, being verified after the rupture of the microstructural mold and the phases formed by scanning electron microscopy (SEM) assays and X-ray diffraction (XRD).

The scheelite residue was collected from an open stock at the Brejuí mine, in Currais Novos, Rio Grande do Norte, Brazil, using only fine waste. Stone dust was obtained from mining in Itapevi, Bahia, obtained by crushing granite rocks. The cassava wastewater was obtained from a flour mill in the municipality of Lagoa Nova - RN, being stored in an open tank for more than 48 h to volatize most of the free cyanide, as verified in the work Silva et al. [15], who considered high loss of toxicity from cassava effluents from 24 to 48 h after being exposed to the environment, and can even be consumed by sheep.

The Portland cement used was CP II F-40 (Portland cement with limestone), which has properties recommended by the Brazilian Portland Cement Association [1] to produce soil-cement compositions. The product was purchased at local stores in the city of Natal, RN, Brazil.

The applied raw materials were selected after which chemical and physical characterizations were carried out. The tests referring to the characterization of the soil in relation to granulometry, NBR 7181 [16], in which the larger particles of 4.8 mm are separated from smaller particles and the sifting is done to know the soil sample retained, making a granulometric curve and density (specific mass) performing dry and wet weighing according to NBR 6508 [17]. For chemical and mineralogical characterization were used SHIMADZU XRD 7000 X-Ray diffraction (XRD) and EDX-720 X-ray fluorescence (XRF) devices. The characterization of cassava wastewater was carried out in several laboratories, starting with the cyanide determination test, acidity tests, hydrogen potential, electrical conductivity, solid and suspension contents, macronutrient and micronutrient tests, chloride, sulfate, and phosphate.

Compaction test was carried out according to standard NBR 7182 [18] to define the formulations and the liquid content of the mixture. In this test, water is added to the solid material to be deposited in a cylinder, being subjected to compaction by a plunger in 26 strokes, in 3 layers. After compaction, a sample is taken for dry weighing, compared to the wet mass. After several tests with an increase in water content, a curve is obtained having an optimum point of greater specific gravity and humidity. After that saturation point, the excess of water promotes a decrease in the specific mass.

Table 1 Compositions of ecological bricks

Nomenclatures	Residues with addition of cement (%)
R6C	Residues + 6% of cement
R12C	Residues + 12% of cement

Note R: Composed of 50% scheelite residue and 50% stone powder; and cassava wastewater; C: Cement

After characterizing the residues, the cement compositions were defined based on the minimum content used for soil stabilization (6%), according to da Silva et al. [5] who used 6–12% cement in their formulations. The contents of solid waste were defined in the packaging analysis as 50% scheelite residue and 50% stone powder concerning the total mass of solid residues. Two compositions were defined for the production of ecological bricks, and their nomenclatures as well as the percentages of each material used in the compositions are shown in Table 1.

The homogenization was performed in a concrete mixer, starting with a mixture of scheelite residue and stone powder to be added to the cassava wastewater and the cement. The press used was the Sahara Hobby brand model for the manufacture of type A bricks (5 cm thick, 10 cm wide, and 20 cm long) according to the NBR 8491 standard, soil-cement brick requirements [19].

Molding and curing followed the precepts of standard NBR 10833 [20], advising that the curing should be with maintaining moisture to hydrate the cement either in a humid chamber or by spraying water. In the case of the work, spraying with cassava wastewater was carried out, avoiding dilution of the same if the brick was moistened with water. Three samples were used for each curing time in each technological test.

The water absorption test was performed in an appropriate tank and dryer, following the Brazilian Standard NBR 8492 [21]. The samples were weighed and placed in a dryer until they reached mass constancy. Soon after, they were weighed again and immersed in water for 24 h, then removed from the immersion and weighed.

The simple compressive strength test (RCS) followed the Brazilian standard NBR 8492 [21], according to this standard the bricks are sectioned in the center and capped (aligned) with cement so that they are level with the contact surface for testing the compression machine.

The durability test was carried out by standard NBR 13554 [22], with the brushing step removed, as according to da Silva et al. [5] the brushing step is not necessary since the requests for surface abrasion for masonry bricks are less stringent than those found for road purposes. In this test, the loss of mass by immersion and drying is evaluated, starting after 7 days of curing. Six cycles were carried out with 5 h of immersion in a tank with water and 42 h of drying in an oven at a temperature of 71 $\pm$ 2 °C, lasting 12 days.

After the rupture of the simple compressive strength test, samples of the formulations with the best results were sent for the analysis of the microstructure of the fracture surface through the SEM and XRD tests. The analysis of XRD peaks was performed using the X'pert tool.

5 Results and Discussion

5.1 Residues Characterization

Regarding the granulometric analysis of the solid residues, the scheelite residue is a fine aggregate with a 0.10 mm fineness module and a maximum diameter of 0.15 mm with spherical particle morphology. The stone powder was characterized as fine aggregate, being consistent with the limits of particle size distribution with a percentage of powdery material less than 10%, with a maximum diameter calculated at 4.8 mm and a fineness module of 2.6 mm with 7.2% powdery material with particle morphology mostly lamellar and rough [23, 24].

Table 2 presents the results of the mineralogical characterization of the residues, highlighting the high calcium content for the scheelite residue found in the DRX in the form of calcite. This characterization was similar to that found in the study by Machado et al. [25], since the waste was obtained from the same mine, indicating the low variability of waste in the region. The stone powder comes from granite rocks, justifying the high silicon content in the quartz phase.

The cassava wastewater stands out for its high potassium content (64.34%) in addition to iron and calcium contents among other chemical elements, as it contains several elements of organic matter and with low crystallinity, and it was not possible to evaluate the crystallographic phases by XRD. About other properties of the cassava wastewater, we have Table 3 in which the total and suspended solids are derived from various compounds contained in the cassava wastewater, especially the starch.

The value of free cyanide found (5.62 mg/L) is even lower than the value obtained in tucupi foods (indigenous food derived from cassava wastewater) according to research by Chisté and de Cohen [26], which was 8.9 mg of HCN/L, that is, it has low toxicity and can be used for manual work. The low pH is due to the presence of acids resulting from the composition of organic matter. The electrical conductivity value of the cassava wastewater was close to the value analyzed in the work of

Table 2 Mineralogical characterization of residues by XRF

Elements	Sheelite residue	Rock dust	Cassava wastewater
Ca	53.37%	4.87%	3.49%
Fe	17.99%	26.99%	18.88%
Si	12.25%	36%	–
Al	4.99%	–	–
K	–	13.59%	64.34%
Zr	–	9.49%	–
Others	11.40%	8.72	13.29
Formed phases	Calcite, quartz, almandine	Quartz, albite, microline	Not applied. Low crystallinity

Table 3 Cassava wastewater characterization

Variable	Valor
Total solids	14,563 mg/L
Suspended solids	2373 mg/L
Free cynide	5.62 mg/L
Total cynide	8.92 mg/L
pH	4.5
Electric conductivity	7.62 mS/cm

Duarte et al. [27] which was 6.8 mS/cm and Barreto et al. [28] which was 7.81 mS/cm, indicating the presence of free ions in the solution that can facilitate cationic exchanges with any material that is added.

The characteristics of the cassava wastewater do not differ much, even in different regions, as verified in the work of Neves et al. [29], Nasu et al. [30] and in the research by Fleck et al. [31] who performed an optimization of the anaerobic treatment of this effluent, obtaining a pH of 4.02 with a 3.2% starch content. According to the authors, the low starch content is important, since its structure is formed by sugars of easy decomposition, generating acids that become inappropriate substrates for the development of methanogenic bacteria. The COD value found was 6014 mg/L and BOD was 1400 mg/L with a ratio between them of 4.2, being considered biodegradable, since it is less than 5 and greater than 2.5 [32].

It is interesting to point out that the organic load of the cassava wastewater, being in an acidic environment, becomes about 150 times more polluting than the organic load contained in a sanitary sewer, being even more serious due to the large volume generated by the industries and because they are concentrated, in some cases close to cities [33].

5.2 *Packing Analysis*

Several factors influence the packaging between particles such as morphology, size, distribution, and compaction technique. The packaging study can be defined as a selection of the proportion and appropriate size of the materials so that the larger voids are filled by smaller particles that will have the voids between them filled by even smaller particles and so on [34].

Scheelite and stone powder residues have different shapes, the first is predominantly spherical and the second is lamellar. After the compaction test, it was found that the voids produced between the lamellar particles of the stone dust were filled by the spherical particles of the scheelite residue, as can be seen in the SEM images in Fig. 1.

Compaction tests were carried out with different proportions between the scheelite residue and stone powder, obtaining the best result for 50% of each residue, since with the proportion of scheelite residue greater than that of stone powder, excess

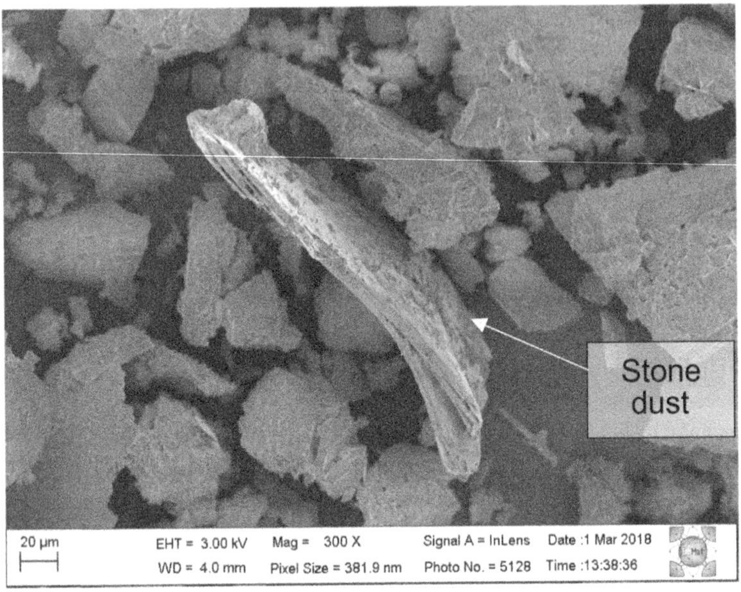

Fig. 1 Stone dust and scheelite residue particles, 300×

according to da Silva et al. [5] in his work on soil-cement brick, the increase in fines content promotes an average increase in the specific surface of the particles present in the mixture, causing an increase in the amount of water necessary for aggregation of the inert particles of the soil. If the amount of stone dust is higher than the amount of scheelite residue, some voids may not be filled between the stone powder particles, generating an increase in porosity, making compaction difficult.

The addition of cassava wastewater to the mixture in substitution to water, when compared to the compaction of residues without the addition of this effluent, promoted a reduction in the optimum humidity, from 11.5 to 11%, keeping the specific mass practically in the same value, as seen in the graph in Fig. 2.

This reduction in moisture content indicates that cassava wastewater promotes better workability in the mixture, better filling the voids between the particles of solid waste as a filler effect, probably due to the high solids content, resulting from the presence of starch as indicated by the works of [35] who studied the effect of cassava starch on the contraction properties of concrete and obtained a decrease in shrinkage (check workability) with the increase in the use of cassava starch by up to 2% by mass, in addition to improved workability.

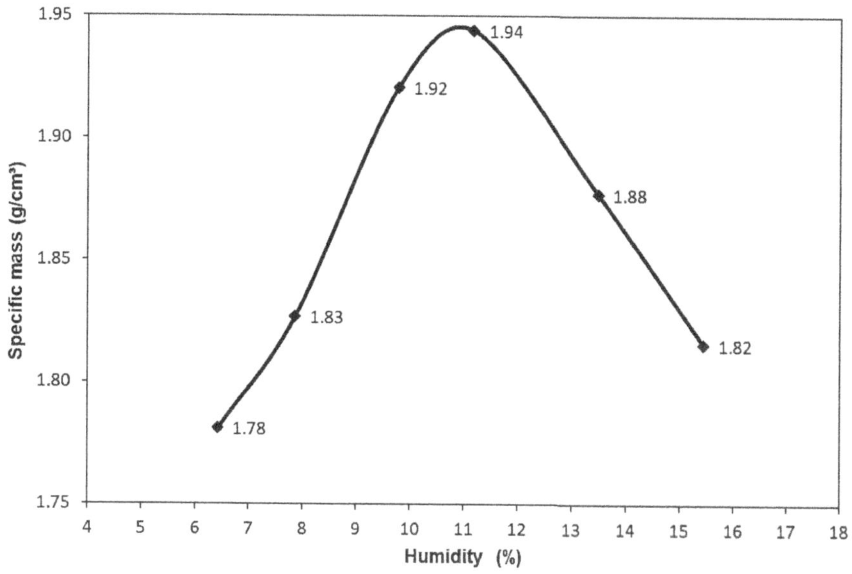

Fig. 2 Compaction chart of the residue mixture

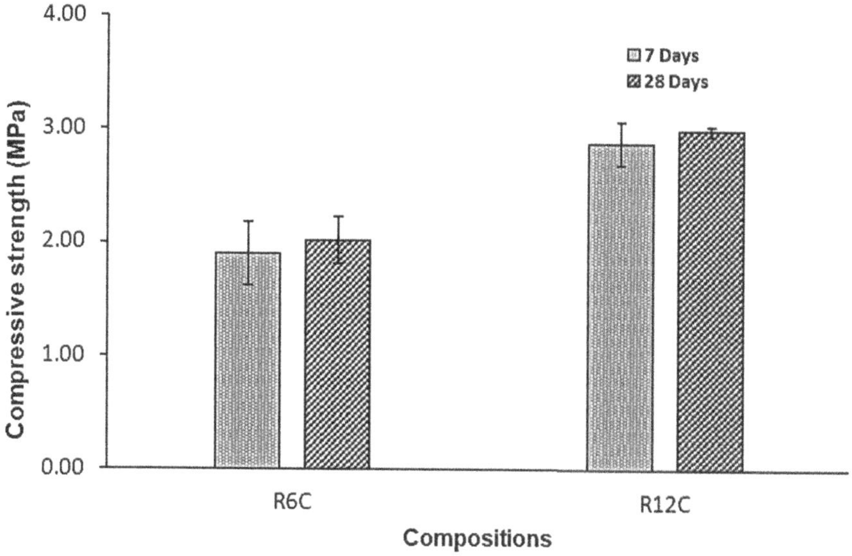

Fig. 3 Simple compressive strength of formulations

5.3 *Simple Compressive Strength*

Figure 3 presents the data of simple compressive strength of the samples of the formulations containing 50% of each solid residue, cassava wastewater, and cement in the contents of 6 and 12% of cement (R6C and R12C) for the curing times of 7 and 28 days. According to [36], in his study on soil stabilization with cement and lime, the minimum percentage of binder recommended for chemical soil stabilization is 6%. This measure can be used as a reference, but the brick is composed only of residues that in this content of 6% of cement added, which corresponds to 5% in the mixture.

Even with a low cement content, the average values of mechanical resistance to simple compression were close to the value recommended by NBR 8491 [19], which is 2 MPa. [9] also found results of simple compressive strength in its bricks having values close to 2 MPa, but considered it suitable for non-structural masonry.

According to the graph in Fig. 8, it can be noted that there was an increase in resistance of the formulation R6C to R12C by approximately 1 MPa, due to the increase in the cement content, promoting more cement reactions. Comparing the curing times of 7 and 28 days, it is observed that there were no significant variations in the formulations, as in the work of da Silva et al. [5] in which there was an increase of approximately 0.5 MPa in the formulation with 12% cement addition from 7 to 28 days, still having the influence of ceramic residue on the pozzolanic effect. It was verified that there are influences promoted by cassava wastewater both in the acceleration and in the delay of the hydration time of the cement.

5.3.1 Delaying Effect of Hydration Time

According to Cody et al. [37], the retarding effect of starches on cement hydration may be due to shielding of portlandite (surfaces which delay their dissolution and consequently the rate of cement hardening reactions. Several authors have related the influence of sugars in retarding cement hydration, such as Smith et al. [38] who evaluated reactions and interactions on the surface of saccharides in cement paste, verifying that adsorption of saccharide species occurs through interactions on the surface of silicates and aluminates. In addition, Kochova et al. [39] evaluated the effect of saccharides on the hydration of Portland cement due to reactions with natural fiber sugars, obtaining a significant impact on cement hydration by slowing hydration for up to 2 days.

According to Kochova et al. [39], sugar decreases the concentration of Ca^{2+} ions in the cement solution, delaying the formation of hydration products. In addition, sugars adsorb on the surface of cement particles, slowing the adsorption rate of H_3O^+ ions, which may inhibit crystal growth and precipitation of C–S–H. High sucrose, K^+ and Na^+ concentrations can form precipitate phases such as syngenite ($K_2Ca(SO_4)2{\cdot}H_2O$) and arcanite (K_2SO_4).

5.3.2 Accelerating Effect of Hydration Time

Singh et al. [40] in his work on the effects of hydroxyethyl cellulose and oxalic acid evaluated changes in cement hydration both in acceleration and in delaying the curing time, evaluating the influence of pH for reactions with cement anhydrous. According to Mehta and Monteiro [41], the accelerating additive should promote the dissolution of the cationic or anionic components of cement, with preference to the dissolution of the constituent that has the lowest dissolution speed during the initial hydration period.

Windt et al. [42] in his experiment on interactions between hydrated cement paste and organic acids, performing thermodynamic modeling of data and speciation evaluated that it is possible that some organic acids strongly accelerated the dissolution of hydrates by acid hydrolysis having effects on cement hydration, and it can even improve mechanical properties.

These acid attacks occur since the contact of the cassava wastewater with the solid residues, especially the scheelite residue, causing the dissolution of $CaCO^3$ calcite and the formation of Ca^{2+} and CO_3^-. The presence of cations in the solution, such as K^+ and Ca^{2+}, both from cassava wastewater and from the dissolution of ions from solid waste tends to promote reactions with silicate and aluminate ions, accelerating the hydration of the cement.

This accelerating effect for the formation of cementitious phases must be more accentuated than the retarding effect promoted by the contents of starch and sugars since the researches regarding the delay in the acceleration of the hydration of the cement were carried out in pure starch and the cassava wastewater is a solution acid that contains only a few starch particles.

It is also possible that some solid particles from the cassava wastewater may be contributing to this effect of accelerating hydration, as occurs in the case of calcite particles that promote acceleration in cement hydration in the early ages by facilitating the nucleation of cement phases, according to research of Irassar et al. [43].

5.4 Water Absorption

It can be seen from the graph in Fig. 4 that all studied compositions showed water absorption values below 20%, established as the maximum limit for soil-cement formulations by the NBR 8491 standard [19]. There is no considerable variation in water absorption between the two formulations, even with an increase in the cement content, since in addition to the densification promoted by cement reactions, there is densification by the chemical reactions promoted by cassava wastewater and by the amount of solute, especially by starch, promoting the filler effect, as verified in the compaction test.

Fig. 4 Water absorption test for formulations

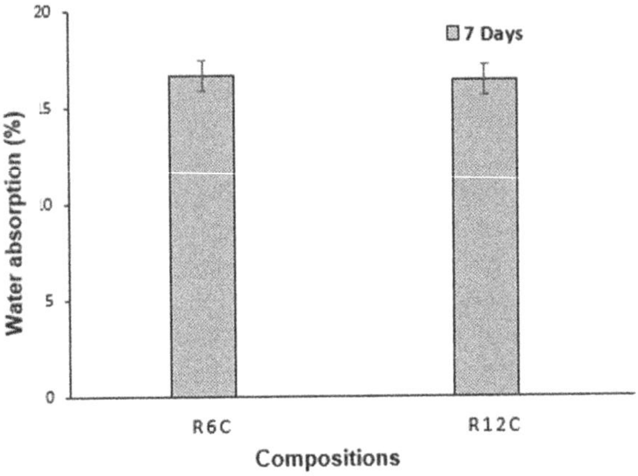

This effect of pore filling by polymeric materials was also verified in the work by Singh et al. [40] on mixtures of mortars modified with polymers, with pores filling by these polymers, obtaining lower values in the water absorption test.

5.5 *Modified Durability*

The results of mass loss obtained values close to those found for soil 12% cement in the works by da Silva et al. [5] which was 0.45% of mass loss, well below the maximum limit of 10% established by NBR 13553 [22], when compared with type A-4 soils.

After the 7 days of the bricks curing, there was a superficial disintegration of the bricks due to the reduced cohesion between the particles, however, the results demonstrate that the brick has durability equivalent to the bricks with clay minerals. It can be said that the cassava wastewater by contributing to a greater adhesion between the particles contributes to durability

In the graph in Fig. 5, it is possible to see the difference in mass loss between the composition with 6% cement addition (R6C) and 12% cement addition (R12C), even with a greater standard deviation, confirming what was observed visually in the bricks, since the bricks of the formulation R6C disintegrated more easily due to the lower content of binder present in the mixture.

5.6 *Fracture Surface Analysis*

Figure 6 shows a micrograph for the formulation R12C for 7 days of curing in which it is possible to observe clusters of particles, several with a spongy aspect, which may be due to calcium oxides and spongy hydrated calcium silicates [44].

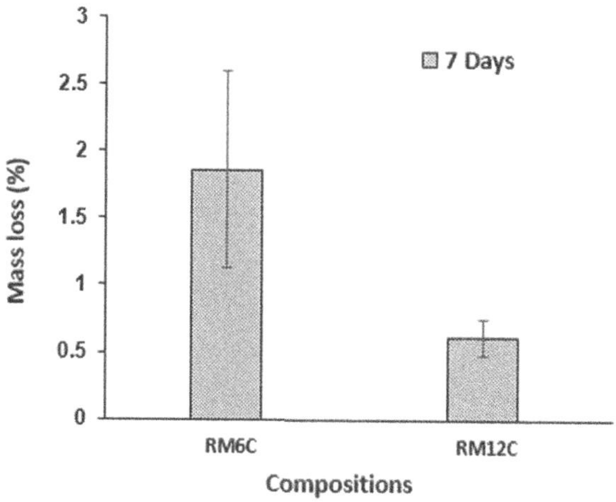

Fig. 5 Formulations loss mass

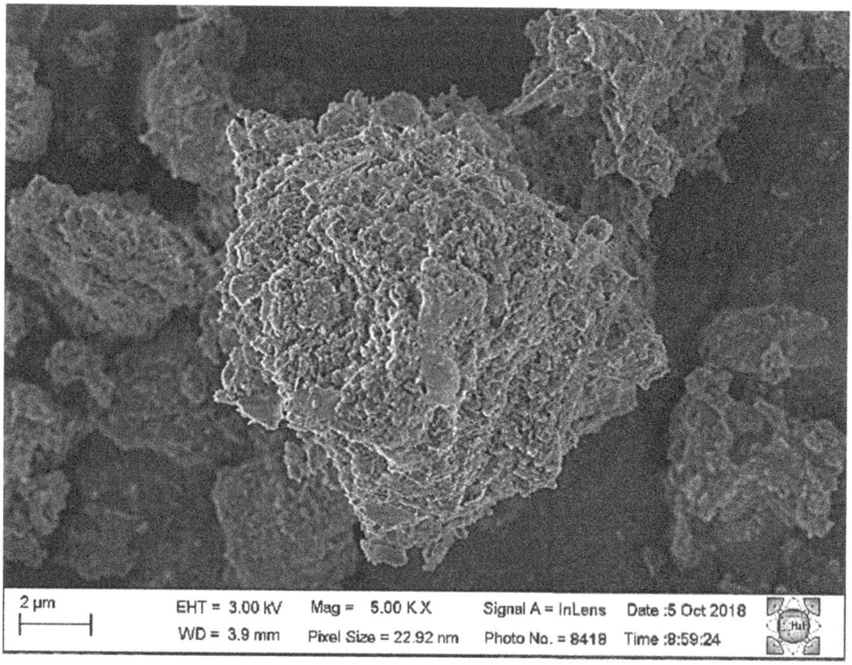

Fig. 6 SEM of the R12C formulation for 7 days of curing, 5000×

In Fig. 7 of the formulation R12C for 7 days of curing, it is possible to observe, in addition to characteristic morphologies of the CSH (calcium silicate hydrated), nanometric globules agglomerated to other particles, more likely to be due to the cassava wastewater, which may be starch nanoparticles, as recently researched in the works of Andrade et al. [45] and Travalini [46]. It is interesting to note that

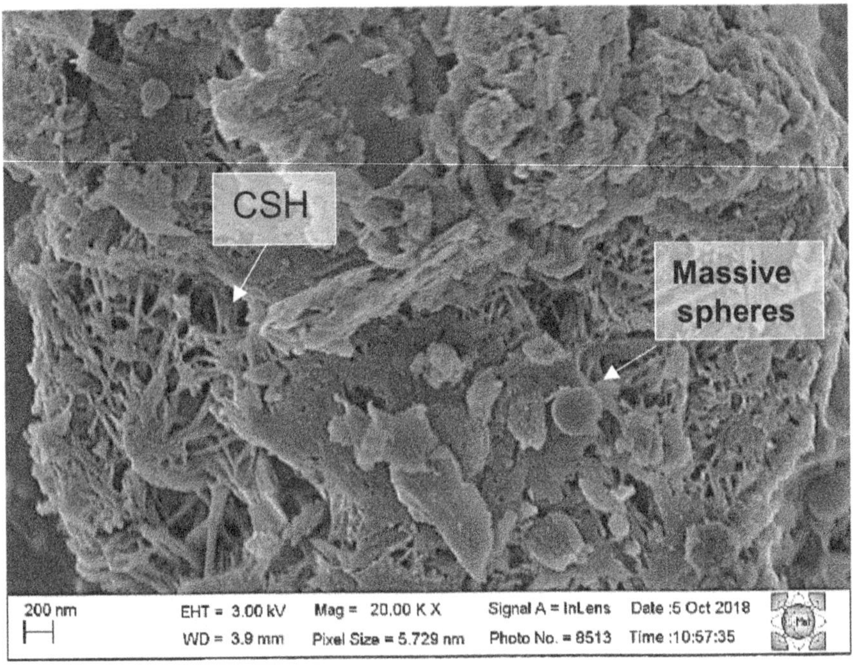

Fig. 7 SEM of the R12C formulation for 7 days of curing, 20,000×

the spherical particles are in contact with the calcium silicates and ettringite, which should be favoring an agglomeration between them and with the waste particles.

Although the starch particles can promote a retarding effect on the hydration of the cement due to the shielding of the portlandite surfaces, according to Cody et al. [37] and Kochova et al. [39], these particles must promote the characteristic plasticizer effect, bringing cement phases closer and after 28 days of curing the cement hardening reactions must have occurred for the most part and the influence of the cassava wastewater acid pH accelerated the hydration reactions cement and residues, facilitating the agglomeration of starch particles.

6 Analysis of the Combination of Solid Residues, Cassava Wastewater, and Cement

Initially, there is a mixture of scheelite residue and rock dust that are considered inert at room temperature. However, there may be a reaction between the oxides present in the residues, such as calcium oxide and moisture forming particle clusters. Then, dry cement and cassava wastewater are added gradually to fully hydrate the mixture .

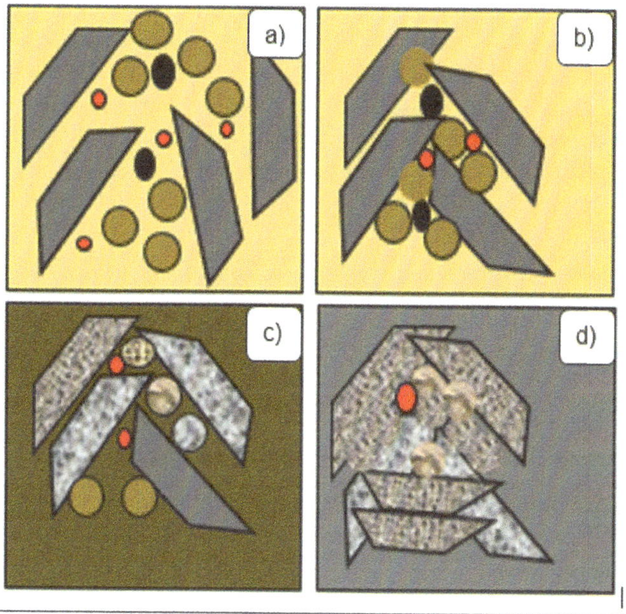

a) Mixing of residues without compaction; b) Mixing after pressing; c) Beginning of cement and chemical reactions with cassava waste water; d) Increase of cement phases with greater compound densification.

Note: Spheres from largest to smallest: scheelite residue, cement and cassava wastewater. Plates: rock dust.

Fig. 8 Schematic representation of the processes in the mixture

Figure 8 presents a schematic representation of the chemical and physical processes that occur in the waste-cement system.

In the pressing process, there is an improvement in the packing between the particles, facilitating the filling of pores (Fig. 8b) and the hydration of the cementitious phases. The action of the cassava wastewater as a binder should also facilitate the precipitation of the cementitious compounds between the waste particles, also occurring the filler effect due to the action of the compounds such as the starch of the cassava wastewater and the finer residues of scheelite and rock dust.

The acids present in the cassava wastewater promote the dissolution of part of the calcite, increasing the amount of $Ca^{2}+$. At the macroscopic level, it is possible to observe the formation of a greater amount of agglomerates, making the mixture more plastic as assessed in the work of Araújo [7]. Chemical reactions between ions are initiated, from waste and dissolved cement particles, calcium, silicates, sulfates (from cement), among other compounds such as calcium carboaluminate hydrated, researched by Ipavec et al. [47].

As seen in Fig. 8c, chemical reactions promote a change in the surface of the particles and some with a change in morphology. These compounds hydrate with the hydroxyl of the water forming portlandite, ettringite, calcium silicates hydrated which must also form other phases with the residues. Along with the curing of the cementitious compound, there is the diffusion of elements and formation of new cementitious phases, densifying and promoting the hardening of the material. Some particles can be seen even after hydrating the brick as seen in the SEM of Fig. 7, being specified in the scheme of Fig. 8d.

7 Conclusions

Based on the analysis of the compaction tests, compressive strength, microstructural analysis, and verification of formed phases, it can be concluded that:

- The high packing degree obtained between solid residues with the addition of cassava wastewater promoted a decrease in porosity and consequently greater mechanical resistance;
- Cassava wastewater contributes to improved packing between particles, provides plasticity to the system, and promotes chemical reactions between residue and cement due to the Ph and its constituents;
- Cement formulations, even with values below the ABCP recommendation (which is 6% for soil-cement stabilization), showed compatible values with current technical standards for soil-cement, along with low water absorption (less than 20%) and resistance to simple compression around 2 MPa;
- Cementitious phases were confirmed by means of SEM, justifying the high mechanical strengths and densities.

These promising results encourage the joint application of these residues in manufacturing bricks, building blocks, and mortars, enabling their rational use in the production chain of the construction industry, in addition to contributing to a reduction of environmental impacts. It is noteworthy that this brick has advantages at the commercial level since the waste is collected *in natura* with no need for further treatment, and it is easily produced and can reduce costs substantially if the bricks are produced in places close to the waste inventories. These market advantages, together with satisfactory scientific results, facilitated the application for an invention patent application, made at National Institute of Intellectual Property (INPI), registered under number BR 10 2019 019143 0.

As the brick produced is composed of fine aggregates, the same composition of solid mineral residues and cassava wastewater can be used for the production of mortars. The study of solid residues with cassava wastewater in the mortar will allow a more detailed analysis of the behavior of this effluent in cement, facilitating the identification of compound properties in the fresh and hardened state, among others, since the cement contents are higher than for bricks.

Acknowledgements This study was financed in part by the Coordenacao de Aperfeicoamento de Pessoal de Nivel Superior—Brasil (CAPES)—Finance code 001, the Federal Institute of Education, Science and Technology of Paraiba and the Federal University of Rio Grande do Norte.

References

1. Brazilian Portland Cement Association—ABCP (2004) Technical Study 35 - Dosage of Soil-Cement Mixtures: Dosing Standards and Test Methods, São Paulo
2. Brazilian Association of Technical Standards—ABNT (2012a) NBR 8491 Soil-cement brick — Requirements, Rio de Janeiro
3. Santos PS (1989) Science and technology of clays, vol 1. Publishing company: Edgard Bucher ltda. SP, Brazil
4. de Ferreira RC, Cunha AHN (2017) Quality evaluation of soil-cement-plant residue bricks by the combination of destructive and non-destructive tests. Rev Bras Eng Agrícola e Ambient 21:543–549
5. da Silva VM et al (2014) Incorporation of ceramic waste into binary and ternary soil-cement formulations for the production of solid bricks. Mater Res 17(2):326–331
6. Faria KCP, Gurgel RF, Holanda JNF (2010) Characterization of sugarcane bagasse ash for use in ceramic bodies. Materials science forum
7. Araújo NC, Ramos AP, Queiroz AJP, dos Santos RC (2015) Mechanical properties of manufactured bricks with soil and cassava wastewater. Rev Verde 10:213–217
8. Acchar W, Marques SKJ (2016) Ecological soil-cement bricks from waste materials [s.i]. Springerbriefs in applied sciences and technology, 64 p
9. Paschoalin Filho JA, Storopoli JH, Dias AJG (2016) Evaluation of Compressive strength water absorption of soil-cement bricks manufactured with addition of PET (Polyethylene Terephthalate) wastes. Acta Scientiarum Technology 38(2):163
10. Brazil (2013) National Department of Mineral Production (DNPM) Mineral economy of brazil. Chapter 8—Aggregates for civil construction, Brazil
11. Bertolino lC, Palermo N, Bertolino AVFA (2012) Geology. In: Aggregates manual for civil construction, 2nd edn. CETEM, Rio de Janeiro, Brazil
12. Machado RV, Andrade FV, Passos RR, Ribeiro RCC, Mendonça ES, Mesquita lF (2016) Characterization of ornamental rock residue and potassium liberation via organic acid application. Brazilian journal of soil Science – Epub
13. Madeira JGF et al (2017) Ecological analysis of hydrogen production via biogas steam reforming from cassava flour processing wastewater. J Cleaner Prod [s.l.] 162:709–716
14. de Carvalho JC et al (2017) Biorefinery integration of microalgae production into cassava processing industry: potential and perspectives. Bioresource Technology [s.l.], pp 1–8
15. Silva VC et al (2017) Experimental poisoning by cassava wastewater in sheep. Pesq Vet Bras. Rio de Janeiro, Brazil 37(11):1241–1246
16. Brazilian Association of Technical Standards—ABNT (2018) NBR 7181, Soil - Grain size analysis, Rio de Janeiro
17. Brazilian Association of Technical Standards—ABNT (2016a) NBR 6508, Gravel grains retained on the 4,8 mm mesh sieve - Determination of the bulk specific gravity, of the apparent specific gravity and of water absorption, Rio de Janeiro
18. Brazilian Association of Technical Standards—ABNT (2016c) NBR 7182. Soil - Compaction test. Rio de Janeiro
19. Brazilian Association of Technical Standards—ABNT (2012c) NBR 108833. Manufacture of brick and block of soil-cement with use of a manual or hydraulic brickmaking machine — Procedure. Rio de Janeiro

20. Brazilian Association of Technical Standards—ABNT (2012d) NBR 13554, Soil-cement — Durability test by wetting and drying — Test method, Rio de Janeiro
21. Brazilian Association of Technical Standards—ABNT (2016b) NBR 8492, Soil-cement brick — Dimensional analysis, compressive strength determination and water absorption — Test method, Rio de Janeiro
22. Brazilian Association of Technical Standards—ABNT (2012e) NBR 13553, Materials for use in monolithic soil-cement wall without structural function - Requirements, Rio de Janeiro
23. Brazilian Association of Technical Standards—ABNT (2011) NBR NM 46, Aggregates - Determination of material finer than 75 um sieve by washing, Rio de Janeiro
24. Brazilian Association of Technical Standards—ABNT (2019) NBR 7211. Aggregates for concrete – Specification. Rio de Janeiro
25. Machado TG, Gomes UU, Monteiro FM, Valcacer SM, Da Silva GG (2012) Analysis of the incorporation of scheelite residue in kaolinitic clay of Boa Saúde-RN. Mater Sci Forum 727–728:844–849
26. Chisté RC, de Cohen KO (2011) Total and free cyanide contents determination during the processing steps for preparing tucupi. Adolfo Lutz Institute Magazine (RIAL). São Paulo, Brazil
27. de Duarte AS et al (2013) Changes in the physical and chemical attributes of a Neossol after application of cassava wastewater doses. Rev Bras Eng Agríc Ambient [online] 17(9)
28. Barreto et al. (2013) Chemical attributes of soil under cassava wastewater application in Marandugrass cultivation. Braz J Agric Sci 8(4):528–534
29. Neves OSC, Souza AS, Costa MA, de Almeida Sousa L, Viana AES, Neves VBF (2014) Persistence of cyanide and ph stabilization in cassava wastewater. Braz J Technol Agroindustrial 8:1274–1284
30. Nasu EGCC, Formentini HM, Furlanetto C (2015) Effect of cassava wastewater on tomato plants infected by the nematode Meloidogyne incognita. Crop Protection 78:193–197
31. Fleck L, Tavares MHF, Eyng E, Andrade MAM, Frare LM (2017) Optimization of anaerobic treatment of cassava processing wastewater. Eng Agríc Jaboticabal 37(3):574–590
32. Loures CCA, Izário Filho HJ, Oliveira IS, Samanamud GRL, Souza AL, Silva MB (2014) The application of response surface methodology in the study of photodegraded industrial dairy effluents by the photo-fenton process: optimization and economic viability. International Scholarly Research Notices, New York, pp 1–9
33. Brazilian Micro and Small Business Support Service, SEBRAE (2015) Primer on cassava waste. DF, Brazil. 2015 ISBN 978-85-7333-596-5
34. Oliveira IRD, Studart AR, Pileggi RG, Pandolfelli VC (2000) Particle dispersion and packaging - principles and applications in ceramic processing. SP, Brazil
35. Akindahunsi AA, Schmidt W (2019) Effect of cassava starch on shrinkage characteristics of concrete. African J Sci Technol Innov Dev 11(4):441–447. https://doi.org/10.1080/20421338.2017.1380580
36. Sherwood PT (1993) Soil stabilization with cement and lime: state of the art review. Hmso, London
37. Cody AM, Lee H, Cody RD, Spry PG (2004) The effects of chemical environment on the nucleation, growth, and stability of ettringite $[Ca_3Al(OH)_6]_2(SO_4)_3 \cdot 26H_2O$. Cem Concr Res 34:869–881
38. Smith BJ, Roberts JR, Funkhouser GP, Gupta V, Chmelka BF (2012) Reactions and surface interactions of saccharides in cement slurries. Langmuir
39. Kochova K, Schollbach K, Gauvin F, Brouwers HIH (2017) Effect of saccharides on the hydration of ordinary portland cement. Constr Build Mater 150:268–275
40. Singh NK, Mishra PC, Singh VK, Narang KK (2003) Effects of hydroxyethyl cellulose and oxalic acid on the properties of cement. Cem Concr Res 33(9):1319–1329
41. Mehta PK, Monteiro PJM (2014) Concrete: structure, properties and materials, 2nd edn. IBRACON, São Paulo
42. Windt l, Berthon A, Larreur-Cayol S, Escadeillas G (2015) Interactions between hydrated cement paste and organic acids: thermodynamic data and speciation modeling. Cement and Concrete Research, Elsevier, vol 69, pp 25–36. Hal-01108814

43. Irassar EF, Bonavetti VL, Menéndez G, Carrasco MF (2015) Hydration and properties of composite cements (with three additions) with limestone and slag. Latin American association for quality control, pathology and construction recovery. ALCONPAT
44. Braz IG, Shinzato MC, Montanheiro TJ et al (2018) Effect of the addition of aluminum recycling waste on the pozzolanic activity of sugarcane bagasse ash and zeolite. Waste Biomass Valorization, pp 3493–3513
45. Andrade P, Camilloto GP, Cruz RS (2018) Production and characterization of breadfruit starch nanoparticles by ultrasonic method. In: 4th Northeast meeting of science and technology of polymers. Aracaju, Brazil
46. Travalini AP (2019) Films of cassava starch reinforced with cellulose nanoparticles for application in "eco-friendly" food packaging. Thesis (PhD in Food Science and Technology), State University of Ponta Grossa, Ponta Grossa, Brazil
47. Ipavec A et al (2010) Carboaluminate phases formation during the hydration of calcite containing portland cement. J Am Ceram Soc

Mortar Produced with Scheelite Residue and Cassava Wastewater

Rayanderson Saraiva de Souza

Abstract This chapter analyzes the feasibility of using scheelite beneficiation residue to replace fine aggregate particles and hydration water by cassava wastewater in mortars. Chemical and mineralogical characterization tests of the materials and an analysis of the mortars in the fresh and hardened state were conducted in seeking to analyze the technical feasibility of using the residues. The results showed better properties in the fresh and hardened state of the mortars due to the presence of the residues.

Keywords Mortar · Residue · Cassava wastewater · Scheelite

1 Introduction

Mortar is compositions of binder, aggregate and water. Aggregates represent 60–80% of the consumption of ready-mixed mortar materials among the components [1]. World aggregate consumption corresponds to about 2/3 of all ore produced on the planet. Brazil's consumption per capita corresponded to 5.7 million tons in 2014, according National Association of Aggregate Producers' Entities for Construction [2]. In addition, the process of obtaining aggregates produces environmental impacts characterized by visual degradation of the landscape, soil, relief, air pollution and groundwater contamination [3].

When analyzing mineral production, Brazil has a high global participation in mineral reserves, with an emphasis on the reserves of natural graphite (31.4%), tantalum (33.5%), manganese (30.1%) and an emphasis on niobium, representing 98.01% of the world reserves, according to National Department of Mineral Production [4]. When analyzing the value of the mineral stake traded, Rio Grande do Norte State, sold approximately R$454,000,000, with the commercialized value of salt

R. S. de Souza (✉)
Digital Metropolis Institute (IMD), Federal University of Rio, Grande do Norte (UFRN), University Campus, Natal, Rio Grande do Norte, Brazil
e-mail: rayanderson_saraiva@hotmail.com

W. Acchar and V. Monteiro da Silva (eds.), *Use of Cassava Wastewater and Scheelite Residues in Ceramic Formulations*, Green Energy and Technology,
https://doi.org/10.1007/978-3-030-58782-6_4

(R$296,800,000), ornamental stones (R$40,856,826) and tungsten (R$30,016,237) standing out according to DNPM data [5].

In turn, tungsten (symbol W on the periodic table) is a metallic chemical element considered one of the rarest elements in the earth's crust, presenting high density of about 19.3 g/cm^3, a high melting point and a high boiling point (5927 °C). It also presents characteristics such as high hardness, resistance to wear and corrosion, good heat and electrical conductivity [6]. The main stratiform deposits of scheelite ore in Brazil are located in the Scheelitífera do Seridó Province in the Seridó region of the states of Rio Grande do Norte and Paraíba in northeastern Brazil, being the largest reserve in the country (DNPM 2016).

In turn, the Brejuí mine stands out as the largest scheelite mine in South America and for processing scheelite ore. According to the mining company, the environmental factors most affected in scheelite extraction are related to the soil (erosions), flora (opening accesses and trails), air (crushing and detonations), gases (pyrite roasting) and aesthetic changes in the landscape (tailings piles) [7, 8].

The scheelite beneficiation process in the mine has approximately 50 m^3 of waste produced daily, approximately 18,000 m^3 of waste/year. In addition, approximately 99.2% of the extracted material is wasted [9], constituting a residue accumulated in two thick and thin fractions in which they are piled up like dunes, with approximately 4.5 million and 2.5 million of coarse and fine fraction, respectively [10].

According to Silva [11], cassava (*Manihot esculenta Crantz*) is a perennial and woody tropical plant adapted to low fertility soils and can be propagated by cuttings or seeds. In turn, cassava wastewater is the milky-looking liquid that drains from the cassava roots which is obtained during the pressing process for obtaining starch or in flour production, therefore being a by-product or even a residue of the cassava processing process, [12]. The inadequate disposal of cassava wastewater generates serious environmental problems such as groundwater contamination. Cassava wastewater also has a high content of hydrocyanic acid and a high biochemical oxygen demand (BOD) and chemical oxygen demand (COD) [13, 14].

Brazil occupied fourth place in the global production of cassava, corresponding to 21.08 million tons in 2015, Food and Agriculture Organization of the United Nations [15]. It is estimated that for each ton of cassava processed in flour houses, 300–600 L of cassava wastewater are generated [16, 17]. In addition, most of the cassava production in Brazil is used for the production of flour, and it is estimated that between 60 and 80% of the total production of cassava roots will be used in the manufacture of the product [18–20]. Some studies have sought to present destination for this residue, as in the production of bio-hydrogen [21–23] and use as a natural insecticide and pesticide [16, 24].

In turn, the replacement of hydration water in the mortar production process enables reducing the high consumption rate of potable water. National Water Agency [25] data show that the transformation and construction industries consume approximately 12% of the total water flows consumed by all sectors of the economy. In addition, potable water can be applied in more noble destinations when it is replaced by cassava wastewater in mortars, which only needs to go through simple filtration to reduce the reduction of organic matter present and exposure to the sun to release some possible cyanides present.

Thus, several studies have been directed to the use of residues in order to replace mortar constituents with residues which contribute to the properties of this material and present an alternative for the disposal of these residues and reduce the consumption of finite natural resources. The use of concrete mixer washing water [26, 27], wastewater [28] and gray water [29] are examples of research that points out the potential of using liquid waste to replace mixing water in concrete and mortar. In addition to this, one can also mention the industrial effluents released from the polyvinyl acetate resin plant that were studied by Ismail and Al-Hashmi [30] when analyzing the production of concretes that replaced the mixing water.

Thus, it was proposed to analyze the properties of these mortars in the fresh and hardened state produced with the substitution of natural fine aggregates by residues from the scheelite beneficiation process and potable water by cassava wastewater, seeking to prove its technical feasibility for use in mortars, and seeking to provide an answer to the destination of the studied waste.

2 Utilized Materials

2.1 Aggregates

This study used medium sand obtained from local stores. This type of sand is commonly used in laying mortar mixes in Brazil.

The scheelite residue used in the formulations came from the Brejuí mine, located in the municipality of Currais Novos/RN, Brazil. Approximately, 70 kg of fine residue and 70 kg of coarse scheelite residue were collected. The collection was carried out in the yards destined for the scheelite residue, and the internal regions of the residue pile (about 80 cm deep) and the central region of the residue amount were used as the sampling area. They were characterized in terms of unitary and specific masses, according to the Mercosur Standard (NM) 45, Brazilian Association of Technical Standards [31]. The X-ray fluorescence tests of the scheelite residue were also carried out, in addition to the mineralogical analysis of the X-ray diffraction, using the Shimadzu XRD-6000 apparatus, with Cu-Kα radiation and scanning angle (2θ) from $5°$ to $55°$, at the Materials Engineering Department of UFRN (Federal University of Rio Grande do Norte).

2.2 Cement

CP II F-40 (Portland cement composed with filler) composite cement was used as binder. This type of cement was chosen due to the ease of acquisition in the city of Natal/RN, in addition to the need to use a cement without the presence of pozzolans, which could interfere in the analysis of possible pozzolanic reactions in the formulations. The cement was stored in a dry place away from moisture and bad weather to preserve its properties. Unit and specific mass tests obtained for natural sand and RS, NM (Mercosul Standard) 45, Brazilian Association of Technical Standards [31] were carried out. In addition to X-ray fluorescence tests, to identify the chemical elements present.

2.3 *Lime*

The lime used was Rebocal CH-II hydrated lime (NBR 7175) [32], manufactured in the municipality of Campina Grande/PB, Brazil. Unit and specific mass tests obtained for natural sand and RS, NM (Mercosul Standard) 45, Brazilian Association of Technical Standards [31] were carried out. In addition to X-ray fluorescence tests, to identify the chemical elements present.

2.4 *Water*

The water used came from the public supply system in the city of Natal/RN, supplied by the local concessionaire, CAERN—Rio Grande do Norte Water and Sewage Company.

2.5 *Cassava Wastewater*

The cassava wastewater was collected from small cassava flour houses located in the city of Lagoa Nova/RN. To decrease the cyanide content present in the manipueira composition, the liquid was exposed to the sun.

3 Methodology

Table 1 presents the features studied in the present work, emphasizing that it refers to the best composition obtained in the granulometric adjustment between the fine and coarse residue when citing scheelite residue. The first mixture in the proportion of 1:3 in volume aimed to analyze the specific behavior of cement hydration in the presence of cassava wastewater and the influence of substituting fine aggregate by the residue. It is emphasized that 1:3 mixtures are widely used in scientific works, as presented by Breitenbach et al. [33]. The second mixture was 1:1:6 (by volume),

Table 1 Mix proportions

Proportions	No	Name	Compositions
1:3 C:S	1	C + SR + CW	Cement + Scheelite residue + Cassava wastewater
	2	C + SR + W	Cement + Scheelite residue + Water
	3	C + S + CW	Cement + Sand + Cassava wastewater
	4	C + S + W	Cement + Sand + Water
1:1:6 C:L:S	5	C + L + SR + CW	Cement + Lime + Scheelite residue + Cassava wastewater
	6	C + L + SR + W	Cement + Lime + Scheelite residue + Water
	7	C + L + S + CW	Cement + Lime + Sand + Cassava wastewater
	8	C + L + S + W	Cement + Lime + Sand + Water

in which the objective was to analyze the same parameters as in the 1:3 mixture, but adding lime as a binder. The 1:1:6 ratio is the mixture most commonly used in the construction industry and in scientific works, as mentioned by Medeiros [7]. Table 1 presents the summary table of the compositions.

3.1 *Molding and Curing of Specimens*

Prismatic specimens were molded in dimensions of 40 × 40 × 160 mm according to NBR 13279 [34], and then they were compacted on the consistency table in two layers. The specimens were demolded after 48 h and kept at room temperature until the test ages of 7, 14 and 28 days, being chosen to keep both mixtures at room temperature since the objective was to analyze the isolated effect of cassava wastewater in the mortar structure, and its curing submerged in water could hinder this analysis.

3.2 *Testing Mortars in the Fresh State*

3.2.1 Mortar Consistency Test

The mortar consistency test was conducted in accordance with NBR 13276 [35], at the UFRN, Construction Materials Laboratory. The test was used to determine the amount of water needed in the compositions, setting the consistency index at 260 ± 5 mm, as recommended by the standard. It is noteworthy that the mortar was tested by a bricklayer in building a kennel for dogs in the UFRN courtyard, which verified its technical application feasibility.

3.2.2 Mass Density and Incorporated Air Content

The test was carried out as described in NBR 13278 [36], conducted at the UFRN Construction Materials Laboratory for all mixtures under analysis.

3.2.3 Water Retention of the Mortars

The test proceeded following the guidelines of NBR 13277 [37] and was conducted at the UFRN Construction Materials Laboratory. This test aimed to understand the mortar's retention capacity, since water retention is an important factor for cement hydration [38].

3.3 Testing of Mortars in the Hardened State

3.3.1 Bulk Density in the Hardened State

Specimens were molded to perform the bulk density test in the hardened state, being performed at 28 days according to the guidelines of NBR 13280 [39], and for each formulation. The test was conducted at the UFRN Construction Materials Laboratory.

3.3.2 Water Absorption by Immersion and Voids Index

The specimens were molded and the procedures described in NBR 9778 [40] were followed for mortar and hardened concrete; determination of water absorption, voids index and specific gravity were performed at 28 days for all formulations.

3.3.3 Capillary Water Absorption and Capillary Coefficient

The test was performed according to NBR 15259 [41] for laying mortar and for covering walls and ceilings; capillary water absorption and the capillary coefficient were determined with the test performed at 28 days using specimens for each formulation. The experimental procedure was carried out at the UFRN Construction Materials Laboratory.

3.3.4 Compressive and Flexural Tensile Strength

The tensile and compression strength tests were performed at the UFRN Building Materials Laboratory according to NBR 13279 [34], using a universal press (Shimadzu, Autograph AG-X model). Four specimens were molded for each age of rupture: 7, 14 and 28 days. The load in the tensile test was applied at a speed of 50 N/s in the center of the specimen with a distance between supports of 100 mm. The specimens were used for the compressive strength test after the tensile rupture, which was subjected to load at a speed of 500 N/s.

3.3.5 Dynamic Modulus of Elasticity

A Pundit Lab + Proceq brand model 32610001 was used to perform this test with 54 kHz transducers and direct reading, performed at the UFRN Cement Laboratory. Specimens were molded for each formulation and the test was performed at 28 days according to NBR 15630 [42] for laying mortar and for covering walls and ceilings; the dynamic elasticity module was determined through ultrasonic wave propagation.

3.3.6 Scanning Electron Microscopy (SEM)

The fracture surfaces of the molded specimens were analyzed to verify the phases formed during the hydration process. Mortar samples were used after breaking for each age and mixture. The tests were carried out in the characterization laboratory of the UFRN Materials Engineering Department using Hitachi TM-3000 equipment. Approximations of up to 20,000× were performed to detect cement hydration products.

3.3.7 X-Ray Diffraction (XRD) of Mortars

Tests were conducted to analyze the hydrated products generated in order to analyze the mineral and chemical composition of the cement paste with cassava wastewater. The test was carried out at the X-ray diffraction and Fluorescence Laboratory at the UFRN Materials Engineering Department, using a Shimadzu XRD-6000 device with Cu-Kα radiation and scanning angle (2θ) from 5° to 55° to identify the mineralogical phases present. The tested samples passed through an ABNT No. 200 sieve (0.075 mm).

4 Results

4.1 Aggregates

The sand size is shown in Fig. 1, presenting: effective diameter (D10): 0.171; non-uniformity coefficient (NUC): 2.977; curvature coefficient (CC): 1.019; and fineness module (FM): 1.97. The sand had a NUC lower than the scheelite residue (SR), which

Fig. 1 Particle size distribution of natural sand and SR

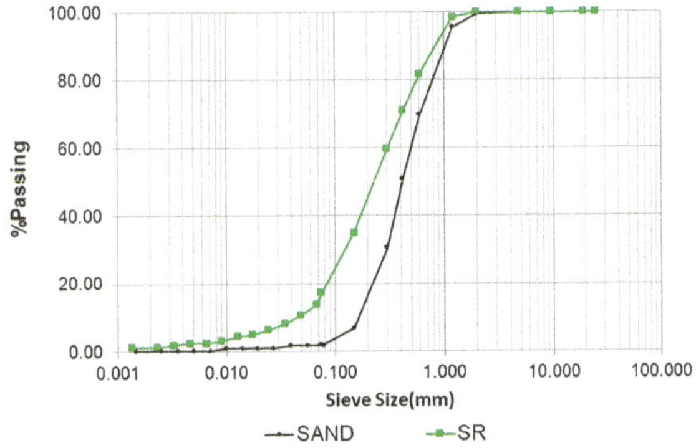

indicates that the sand particles are more uniform, corroborating for greater void formation and consequently lower density which can generate lower performance properties such as the greater presence of voids and a mortar with lower density Brazilian Standard [35] 7181.

The results of the granulometric adjustment in which it was sought to improve the packing of the particles are shown in Fig. 1, where it can be seen that the thick fraction of the residue has a real density of 2.92 g/cm^3 and the fine fraction of the residue presents 2.91 g/cm$^{3.}$

The curve also shows the progressive increase in specific gravity between the adjustment with 0% of the fine residue and 100% of the thick residue, to the point of greatest density in the adjustment with 25% of the fine residue and 75% of the thick residue, along with the specific gravity of 3.04 g/cm^3 and unit mass of 1.74 g/cm^3. The specific gravity decreases after that point as more particles of the coarse fraction are added.

Thus, using the verified results, 25% of the fine fraction and 75% of the coarse fraction of the scheelite residue (now referred to as SR) were defined for all analyzed formulations, since the higher density in the aggregate significantly contributes to obtaining better performing properties in the analyzed study, as the aggregate can correspond to up to about 80% of the mortar volume [1].

Figure 2 shows the granulometric curve obtained for the adjustment with 25% of the fine residue and 75% of the coarse residue, while Fig. 2 shows the curve obtained for the medium sand in order to perceive the adjustment in the fines made when using the fine scheelite residue. It is possible to perceive continuous distribution of the SR in the curve, presenting effective diameter (D10): 0.046; non-uniformity coefficient (NUC): 6.631; curvature coefficient (CC): 1.207; and fineness module (FM): 1.25. The granulometric distribution corroborates with the result regarding the specific gravity for this composition, since the continuous distribution of the particles allows the best accommodation of the particles and consequent reduction of voids [43].

Table 2 condenses the unitary and specific densities obtained for natural sand and SR, Brazilian Association of Technical Standards [31] NM 45.

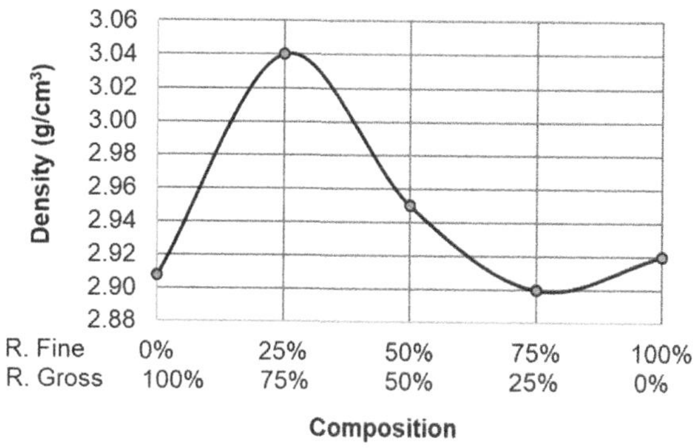

Fig. 2 Density for different compositions with fine aggregate and gross aggregate of the scheelite residue

Table 2 Dry density and real density of the natural sand and SR

Aggregate	Dry density (g/cm^3)	Real density (g/cm^3)
SR	1.74	3.04
Natural sand	1.45	2.65

Table 3 Dry density and real density of lime and cement

Binder	Dry density (g/cm^3)	Real density (g/cm^3)
Cement	0.98	3.08
Lime	0.82	2.62

4.2 Binders

Table 3 presents the summary of unit and real density obtained for cement and lime, NBR 16605 [44].

4.3 Cassava Wastewater

The results obtained in the characterization tests of the liquid cassava wastewater are presented in Table 4. The laboratory results obtained from cassava wastewater indicate high potassium, nitrogen, sodium, iron and calcium concentrations, presenting similar values to those presented in the literature by Neves et al. [45] and Nasu et al. [46]. Variations in the contents are due to factors such as climate, soil and type of crop, in addition to the way the waste is stored.

The pH of the sample was classified as acidic (4.5), as well as the average of the authors' values presented in Table 4 (4.35). However, acidity did not compromise the mortar properties when compared to mortars without using cassava wastewater. The presence of total solids in the sample (14,563 mg/L) deserves attention, since

Table 4 Chemical composition cassava wastewater

Components	Values
Total nitrogen (mg/L)	1121
Phosphorus (mg/L)	132
Potassium (mg/L)	1456
Calcium (mg/L)	93
Magnesium (mg/L)	219
Sulfur (mg/L)	74
Iron (mg/L)	117
Zinc (mg/L)	20
Manganese (mg/L)	15
Copper (mg/L)	9
Sodium (mg/L)	351
Chloride (mg/L)	545.4
SST: total suspended solids (mg/L)	14,563
Suspended solids (mg/L)	2373
pH	4.5
Free cyanide (mg/L)	5.62
Total cyanide (mg/L)	8.92
Electrical conductivity (mS/cm)	7.62

the presence of solids had an influence on the mortar properties such as immersion absorption, capillary and resistance to compression and traction.

The cassava wastewater also presented low total cyanide content (8.92 mg/L), below the average presented by the authors [13, 14]. In addition, the free cyanide content (5.62 mg/L) was within the limit established by NR 15 [47], which establishes a maximum limit of 8 mg/L in work environments with a workload of up to 8 h per day.

4.4 *Chemical and Mineralogical Analysis of Materials*

The results obtained by X-ray fluorescence are summarized in Table 5, in which it is possible to verify the highest concentrations of CaO, Fe_2O_3, SiO_2 and Al_2O_3 elements commonly found in the literature [48]. Lime, cassava wastewater and scheelite beneficiation residue show higher CaO concentrations. The results from the processing residue corroborate those presented by Paiva [10], Gerab [9] and Medeiros [7]. Likewise, the peaks shown in the X-ray diffractogram (Fig. 3) corroborate with the same authors, with higher peaks for $CaCo_3$, quartz (SiO_2) and grossular (Ca_3Al_2 $(SiO_4)_3$). Therefore, the high presence of CaO in cassava wastewater and SR could contribute to better plasticity of mortars, as shown in the mortar consistency test.

Table 5 X-ray fluorescence of the cement, lime and cassava wastewater

	CaO	Fe_2O_3	SiO_2	Al_2O_3	K_2O	SO_3	P_2O_5	Na_2O	SrO	MgO
Cement	83.16	8.70	6.54	1.60	–	–	–	–	–	–
Lime	96.91	1.39	1.19	0.51	–	–	–	–	–	–
Cassava wastewater	25.34	15.92	2.46	2.31	32.65	3.12	3.59	9.99	–	–
Scheelite residue	58.51	15.51	11.68	4.16	–	1.35	–	–	2.95	2.15

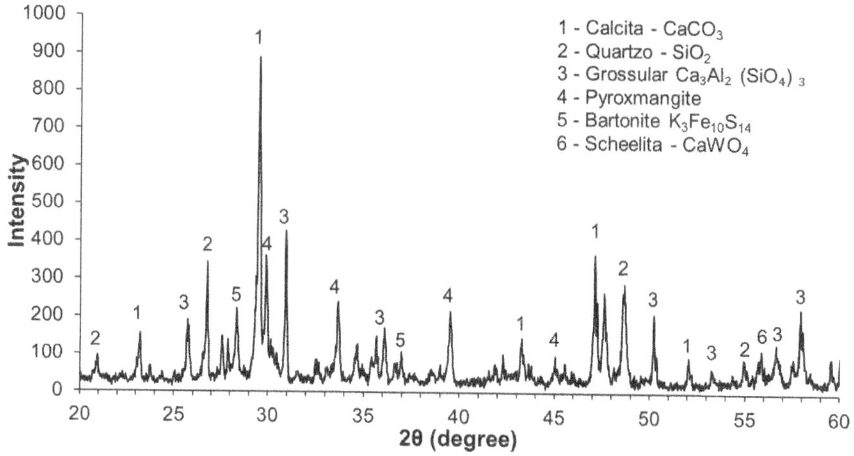

Fig. 3 X-ray diffraction of the residue of scheelite

4.5 *Testing Mortars in the Fresh State*

4.5.1 Mortar Consistency Test

Table 6 shows the ratio (water/cassava wastewater) of dry materials/water or cassava wastewater obtained in determining the consistency index for the 260 ± 5 mm spread. In analyzing each residue in isolation, it is first perceived that the presence of SR in the compositions led to a reduction in the w/c ratio, meaning a decrease in the need for water/cassava wastewater, corroborating the results presented by Medeiros [7] and Neto [49]. In addition, SR has high levels of CaO, which contributes to the increase in plasticity [9, 10, 49].

The contribution to reducing the w/c ratio for all formulations is perceived when analyzing the substitution of hydration water for cassava wastewater. Volatile chemical elements are constituents of cassava wastewater [50], and the mechanical homogenization process stimulates the release of these elements, making the mortar possess a more aerated aspect, thus incorporating more air [51] which favors greater mortar fluidity. In addition, the cassava wastewater composition contains starches

Table 6 Water/cassava wastewater ratio

	Name	A/c
1:3	C + SR + CW	0.1475
	C + SR + W	0.1524
	C + S + CW	0.1466
	C + S + W	0.1736
1:1:6	C + L + SR + CW	0.1567
	C + L + SR + W	0.1747
	C + L + S + CW	0.1610
	C + L + S + W	0.1962

Note *C* cement; *L* lime; *S* natural sand; *SR* scheelite residue 75% gross and 25% fine; *W* water; *CW* cassava wastewater; *a* water/cassava wastewater (g); *c* particles

and biopolymers which act as a plasticizer additive, resulting in the dispersion of particles and contributing to increase the mortar fluidity [52–55].

The same behavior mentioned was noticeable for both analyzed compositions; however, the presence of lime in a 1:1:6 ratio increased the w/c ratio, which is indicated by the high specific gravity of lime and cement, in addition to their chemical composition [1, 7, 38, 56].

4.5.2 Wet Density and Incorporated Air Content

The result of wet density in the fresh state is shown in Fig. 4a for the C + SR + CW, C + SR + W, C + S + CW and C + S + W mixtures, and in Fig. 4b for the C + L + SR + CW, C + L + SR + W, C + L + S + CW and C + L + S + W mixtures. The presence of cassava wastewater in the formulations contributed to a mortar with a lower mass density when compared to mortars which used water. The result corroborates with the consistency test with respect to cassava wastewater working as an air-incorporating additive, contributing to a lower density in the fresh state [50].

In analyzing the substitution of SR for natural aggregate, Fig. 4a and b shows the highest mass density in the fresh state for formulations with the presence of SR. The behavior is expected because the specific gravity of SR is greater than the specific gravity of the natural aggregate [7, 9, 10].

Choi et al. [57] used high-density glass waste to replace the natural aggregate and obtained mortars with higher density than mortars with natural aggregate, being justified by the high density of the glass when compared to the natural aggregate. The same behavior is indicated by Santamaría-Vicario et al. [58] when replacing the natural fine aggregate with steel slag.

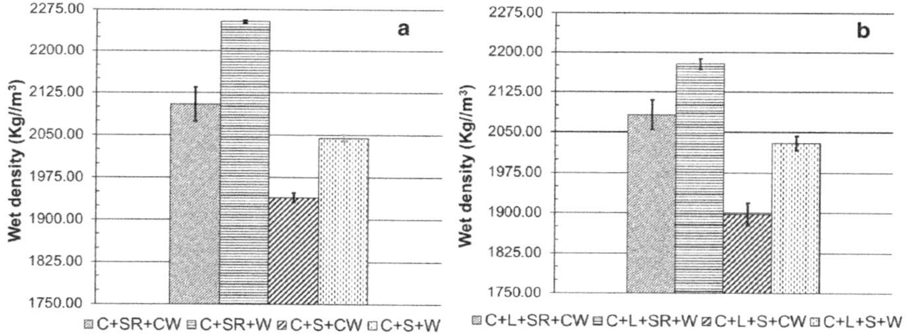

Note: C — cement; L- lime; S — Natural Sand; SR — Scheelite residue 75% gross and 25% fine; W — water; CW - Cassava wastewater.

Fig. 4 Wet density for the proportions 1:3 (**a**) and 1:1:6 (**b**)

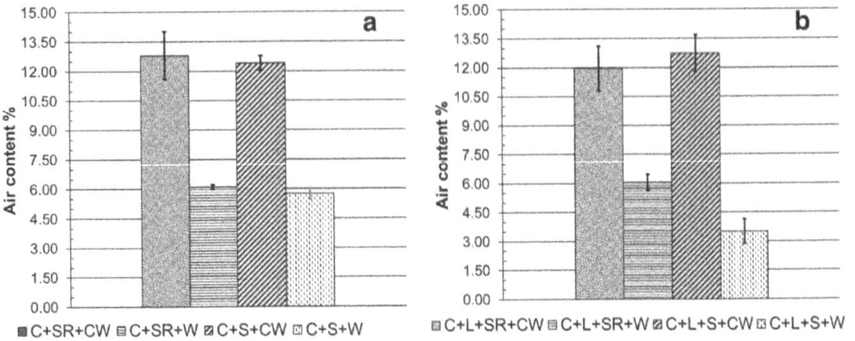

Note: C — cement; L- lime; S — Natural Sand; SR — Scheelite residue 75% gross and 25% fine; W — water; CW - Cassava wastewater.

Fig. 5 Air content for the formulations 1:3 (**a**) and 1:1:6 (**b**)

The presence of lime in the compositions developed mortars with a slightly lower tendency for mass densities in the fresh state than mortars without lime, also associated with the specific gravity of lime (2.62 g/cm^3) being lower than that of cement (3.08 g/cm^3).

The result of the incorporated air content for the compositions is presented in Fig. 5a and b. The result reinforces the air-incorporating behavior presented by cassava wastewater in the consistency and mass density test. Compositions with cassava wastewater showed about four times more incorporated air than compositions without cassava wastewater.

Romano [51] states that incorporating air tends to improve the cohesion of the coating mortars, reduces the tendency to exudate, improves plasticity and makes it possible to reduce the amount of water needed for kneading, indicating a possible decrease in plastic shrinkage and drying. Likewise, the incorporation of air increases the separation between the aggregates, functioning as a kind of lubricant for the particles, contributing to the mixture and flow. Romano [51] indicates the existence of a critical point when analyzing the mixing time, in which a maximum content of incorporated air is obtained and then little increase is noticed after that point.

4.5.3 Mortar Water Retention

The result of the water retention test is shown in Fig. 6a and b, where it is possible to observe that the presence of cassava wastewater in the mortars replacing the hydration water showed a tendency to retain less water, with the difference being more noticeable in the proportions of 1:1:6.

When using natural sand as aggregate in the formulations, the presence of lime favored the tendency to greater retention (C + L + SR + CW, C + L + SR + W) when compared to formulations with cement only (C + SR + CW, C + SR + W. The substitution of the aggregate for SR presented a tendency for less retention, being

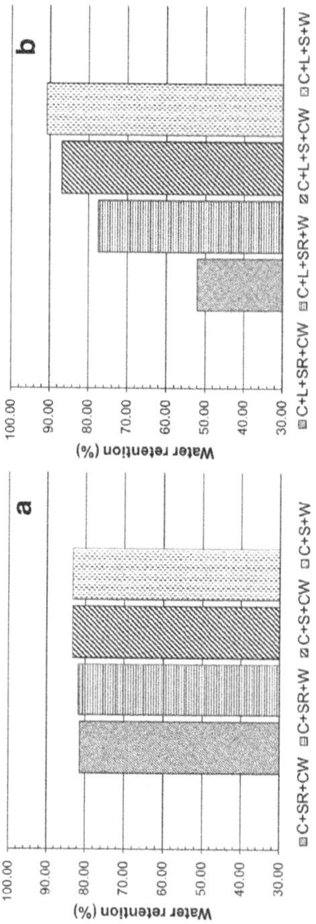

Note: C — cement; L- lime; S — Natural Sand; SR — Scheelite residue 75% gross and 25% fine; W — water; CW - Cassava wastewater.

Fig. 6 Water retention for the formulations 1:3 (**a**) and 1:1:6 (**b**)

indicated by the shape of the SR particles not having a porous texture, which reduces their retention capacity [9, 10, 56].

Mortars with a high retention capacity tend to lose the mixing water more slowly, being progressively displaced to the strength gain will significantly contribute to the appearance of cracks. In addition, water retention is an important factor for developing good workability and adhesion of the mortar to the substrate [38, 59].

4.6 *Testing Mortars in the Hardened State*

4.6.1 Bulk Density in the Hardened State

The results obtained for formulations in the proportion of 1:3 for the density test in the hardened state are presented in Fig. 7a. When analyzing the presence of SR, it is noticed that the formulations which used SR showed higher densities, 2035.38 ± 22.40 kg/m^3 and 2112.25 ± 19.30 kg/m^3 (C + SR + CW and C + SR + W formulations, respectively) when compared to the formulations which used the natural aggregate. This result was expected due to the granulometric adjustment of the residue, obtaining a higher specific gravity than the value of the natural aggregate. The same principle is presented by Paiva [10], Gerab [9] and Medeiros [7].

A lower density was found for formulations which replaced conventional hydration water with cassava wastewater when analyzing the presence of cassava wastewater in the compositions. In addition, the cassava wastewater presented an air-incorporating behavior, which contributed to the lower densities in the compositions which replaced the hydrating water with the cassava wastewater, in addition to the contribution of the dispersing effect caused by the presence of the cassava wastewater starch [52–55].

Figure 7b presents the density result in the hardened state for the proportions of 1:1:6. The compositions showed the same trends which occurred for the 1:3 ratio, namely higher density for formulations with SR and lower densities when replacing hydration water with cassava wastewater. However, the densities of the 1:1:6 ratio were even lower than the 1:3 ratio, being associated with the presence of lime which has a lower specific gravity (2.62 g/cm^3) than cement, generating smaller product densities than produced by hydrating the cement [1, 38].

Baali et al. [43] performed the ternary composition with siliceous sand (SiS), slag sand (SlS) and brick residue sand (BRS) to replace the natural fine aggregate. The results showed improvements in properties, especially due to the better packing of the particles. Similarly, Choi et al. [57] obtained the same trend when replacing the natural aggregate in 25, 50, 75 and 100% with high-density glass associated with high specific gravity of the residue used.

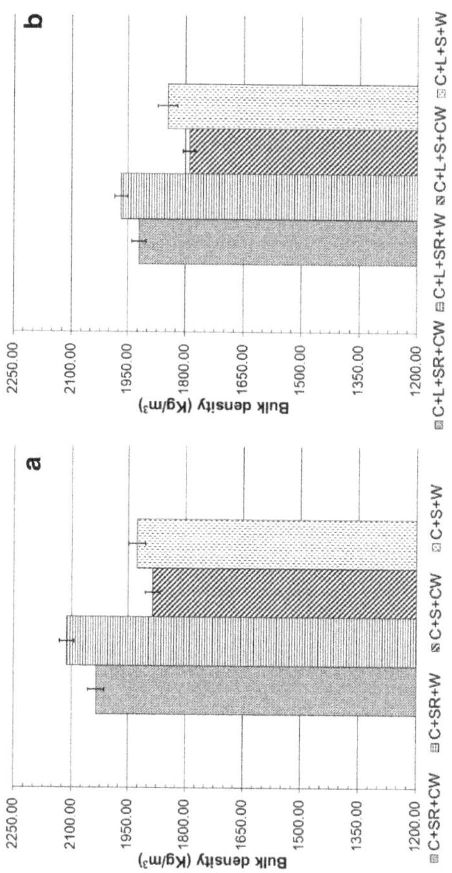

Note: C — cement; L- lime; S — Natural Sand; SR — Scheelite residue 75% gross and 25% fine; W — water; CW - Cassava wastewater.

Fig. 7 Bulk density for the proportions 1:3 (**a**) and 1:1:6 (**b**)

4.6.2 Water Absorption and Voids Index

The results obtained in the water absorption test for the proportions of 1:3 and 1:1:6 are summarized in Fig. 8a and b, respectively. In the results, it was noticed that the mortars with lime showed superior absorption to the formulations without lime. Compositions without lime showed absorption between 14 and 18%, while they presented variations between 12 and 15% in formulations without lime. In addition, there is a tendency for less absorption in mortars with the use of cassava wastewater, attributing the presence of total solid particles of cassava present in the cassava wastewater which may be functioning as a filler in the structure, leading to possible pore refinement, and also the presence of starch which is a biopolymer and can corroborate for less absorption [52, 53, 55, 60].

The results obtained in determining the voids index for the proportions of 1:3 and 1:1:6 are also presented in Fig. 8a and b, respectively. Lime formulations showed higher voids (between 29 and 31%) than mortars which only used cement (between 23 and 27.5%) as a binder.

In analyzing the two factors together, it is observed that there was a tendency of less absorption in the compositions with cassava wastewater in the proportions of 1:3, contributing to the voids index result (Fig. 8a) of the 1:3 ratio in which a tendency of a lower voids index occurred in mortars with the presence of cassava wastewater. In observing the 1:1:6 compositions, we can see a tendency of greater absorption in the mortars with cassava wastewater, also corroborating the higher voids index shown in Fig. 8b.

When analyzing the effect of the presence of only SR in the compositions, it was noticed that substituting the natural aggregate for SR led to a reduction in absorption and voids index in all formulations. Despite having a higher incorporated air content in the fresh state in mortars with the presence of cassava wastewater, the reverse occurred in the void index. During the experiment, it was observed that bubbles in the mortar were released even after the mortar molding process, being attributed to the volatile biochemical elements present in the cassava wastewater which may have contributed to the fact that it had a lower void index in the hardened state.

In addition, two factors may have contributed to the volatilization of gases. The first concerns the fact that the test of incorporated air content is carried out using the mortar right after mechanical mixing with a certain incorporation being presented, then later when the mortar leaves for molding, the specimens go through the thickening process which makes more air come out; this may be related to the lower voids index in the hardened state. Another reason that can be listed is the presence of total solids and starch (biopolymer) in the structure, which may be causing a filler effect in the structure, reducing the voids.

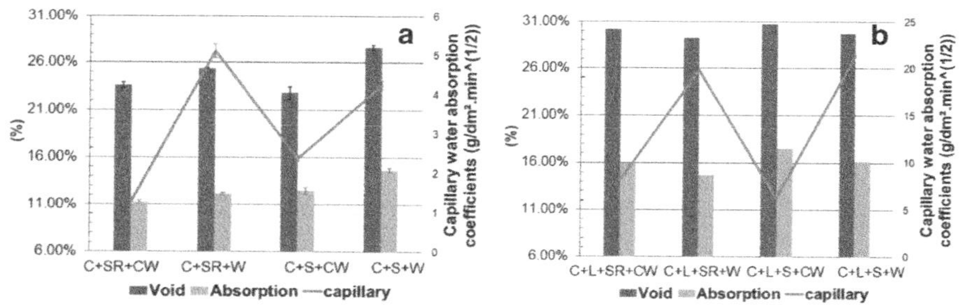

Note: C — cement; L- lime; S — Natural Sand; SR — Scheelite residue 75% gross and 25% fine; W — water; CW - Cassava wastewater.

Fig. 8 Water absorption, water capillary absorption, and voids for the proportions 1:3 (**a**) and 1:1:6 (**b**)

Table 7 Water capillary absorption: 10 min (A10), 90 min (A90) e capillary water absorption coefficients

Proportions	Name	A10	A90	Capillary water absorption coefficients (g/dm^2 min$^{(1/2)}$)
1:3	C + SR + CW	0.11	0.19	0.08
	C + SR + W	0.31	0.63	0.32
	C + S + CW	0.13	0.28	0.15
	C + S + W	0.26	0.52	0.26
1:1:6	C + L + SR + CW	0.44	0.95	0.51
	C + L + SR + W	0.74	1.98	1.24
	C + L + S + CW	0.13	1.10	0.98
	C + L + S + W	0.96	2.30	1.34

Note *C* cement; *L* lime; *S* natural sand; *SR* scheelite residue 75% gross and 25% fine; *W* water; *CW* cassava wastewater

4.6.3 Capillary Absorption and Capillary Coefficient

Capillary absorption at 10 min (A10) and 90 min (A90) is summarized in Table 7, as well as the capillary coefficient. The results show the highest capillary absorption for formulations without cassava wastewater, meaning that the use of cassava wastewater contributed to reducing capillary absorption. Compositions with cement only obtained lower absorptions than compositions which used lime, corroborating the result presented in the voids index.

In addition, it was noticed that the result shows low capillary rise in the samples with cassava wastewater, being an indication of the little pore connectivity present in the samples. The results compete with the effect mentioned above, in which the cassava particles present in the cassava wastewater may be functioning as a filler in the sample structure, providing pore refinement, hindering capillary growth, and the presence of starch can additionally be contributing to sealing existing voids in the microstructure and preventing or hindering the capillary rise of water [48, 61–63].

Less capillary absorption is also perceived in mortars with the isolated presence of SR as a substitute for fine aggregate, corroborating the void index test result and the granulometric adjustment performed.

4.6.4 Compressive Strength

Figure 9a and b shows the results of the compressive strengths at 7, 14 and 28 days for the 1:3 and 1:1:6 mixtures, respectively. The progressive increase in strength over the days is noticeable in the figures. The C + SR + CW composition showed the highest strength in the 1:3 mix, obtaining 17.40 ± 0.31 MPa at 28 days. It is also noticed that the compositions with scheelite or cassava wastewater, used in isolation or together, obtained better strength when compared to the reference mortar (C + S

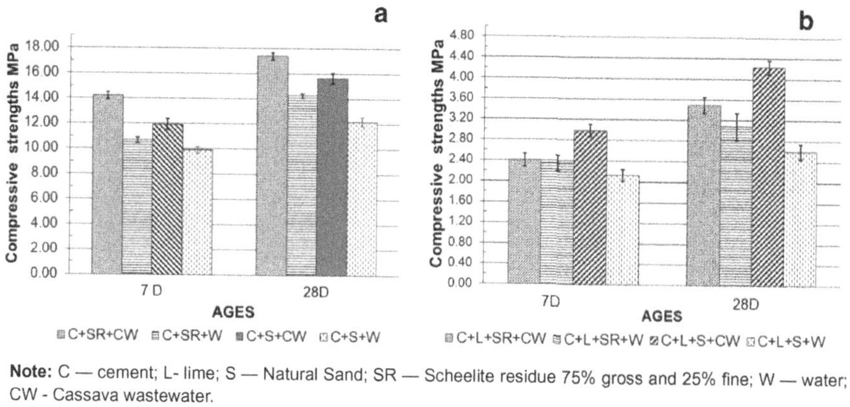

Fig. 9 Compressive strength for the proportions 1:3 (**a**) and 1:1:6 (**b**) with 7, 14 and 28 days

+ W). The greatest strength gain in the proportions of 1:3 occurred in seven days, representing on average 75% of the strength obtained. The same behavior occurred in the 1:1:6 compositions, being represented by about 68% of the compressive strength obtained at 28 days.

Thus, we can summarize that the compressive strength results are associated with three factors, with the first concerning the granulometric adjustment carried out on the scheelite residue, which provided a more compact structure of the fine aggregate and which enabled a denser mortar, favoring the mechanical strength [1, 7, 10].

Baali et al. [43] performed the granulometric adjustment using silica sand, slag sand and brick residue sand. The ternary composition which best contributed to the mortar properties was the composition with 70% silica sand, 15% slag sand and 15% brick residue sand, achieving higher strengths than the reference composition.

The second factor is pointed out in the XRD of the samples, in which it is possible to perceive peaks with similar intensity between the C + SR + H and C + SR + CW samples, however, with the tendency to show more intense CSH peaks in the sample with cassava wastewater; this indicates its greater hydration, thereby contributing to the development of better strength due to the fact that CSH is the main factor responsible for strength [1, 48, 64, 65].

The third factor is the presence of cassava wastewater. The behavior of the mortar in the cassava wastewater incorporated air into the composition, which in the first moment would lead to expect lower strength; however, the cassava wastewater presents starch and sugars in its composition which contribute to its functioning as a biopolymer, which in turn can form a component when drying that contributes to the mortar strength [52, 53, 55, 62, 66]. In addition, cassava wastewater presents small particles derived from the cassava pressing process, namely total solids (14,563 mg/L), which may be acting as a filler in the mortar structure, contributing to pore refinement and a consequent increase in the mortar strength, as occurred in capillary absorption.

The 1:3 formulations showed higher strength than the 1:1:6 mixture, with this being an expected result due to the presence of the lime in which the generated product presents lower strength than the product generated in cement hydration [48]. This fact may be related to the greater solubility of calcium hydroxide, as well as the flocculation of SR particles which contribute to increase the voids in the formulations when added to lime.

4.6.5 Tensile Strength

The tensile strength results of mortars showed the same behavior observed for compressive strength and are presented in Fig. 10a and b for the proportions of 1:3 and 1:1:6, respectively. The tensile strength in formulations with the presence from SR and cassava wastewater showed higher values with better results at 28 days for the C + SR + CW composition with 4.90 ± 0.16 MPa, and this behavior is perceived for all formulations [1, 67]. These results corroborate with Medeiros [7], who obtained a gain in tensile strength when using the coarse fraction of the scheelite beneficiation residue. The tensile strength in the samples containing cassava wastewater reinforces the indicated action of the starch present in the cassava wastewater to act as a system which seals the existing pores, contributing to increase the strength value of mortars. The formulations showed lower strength values in mortars with the presence of lime, as occurred in the compressive strength. The greatest strength gains for all compositions were obtained at 7 days, representing about 70% of the tensile strength at 28 days.

This behavior corroborated that presented in the compressive strength, in which the samples with the presence of SR and cassava wastewater had higher strengths and the samples with the presence of cassava wastewater showed greater tensile strength, despite the lower density. Thus, the higher density acquired through the

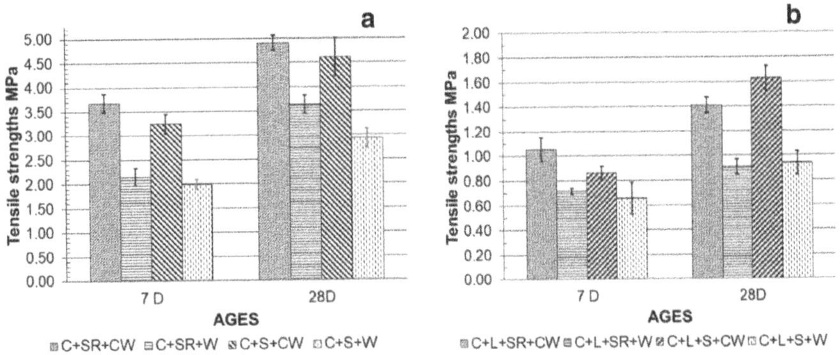

Note: C — cement; L- lime; S — Natural Sand; SR — Scheelite residue 75% gross and 25% fine; W — water; CW - Cassava wastewater.

Fig. 10 Tensile strengths for the proportions 1:3 (**a**) and 1:1:6 (**b**) with 7, 14 and 28 days

granulometric adjustment in SR corroborated the strength gain, but the presence of cassava wastewater is also determinant for this increase in strength.

The strengths followed the same trend in the comparative analysis between the compressive and tensile strength results: the greater the compressive strength, the greater the tensile strength tends to be, as presented in the literature [64, 67, 68].

4.6.6 X-Ray Diffraction and Scanning Electron Microscopy (SEM)

A diffractogram for C + SR + W and C + SR + CW samples at 28 days is expressed in Figs. 11 and 12, respectively, in which it is possible to perceive etringite, portlandite, hydrated calcium aluminate silicate (CAS), hydrated calcium silicate (CSH), calcite, silica, dicalcium silicate (C_2S) and tricalcium silicate (C_3S), peaks, which were expected for products from anhydrous cement as well as from its hydration process.

The peaks between the two samples were similar in intensity and mineralogy. However, there is the appearance of CSH and CAS peaks with greater intensity in the C + SR + CW composition than in the formulation without cassava wastewater, such as the peak close to 21 (2σ). de Souza [56] performed analyzes on soil–cement compositions replacing the hydration water with cassava wastewater, noting the increase in the strength of the solid bricks, albeit in a subtle way, justifying the possible increase in the induced cation exchange capacity due to the cassava wastewater acidity. The behavior is also indicated by Barreto et al. [69], who perceived the increase of this cation exchange capacity when using cassava wastewater in soils for agronomic cultivation, also increasing the amount of interchangeable calcium.

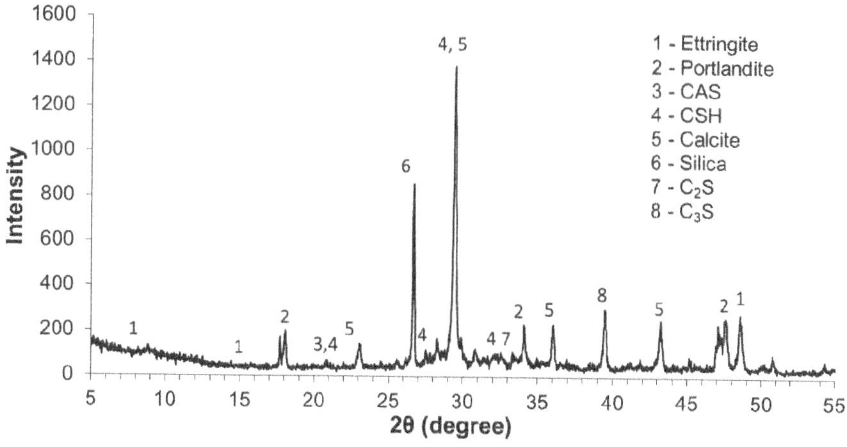

Fig. 11 XRD analysis—formulation C + SR + W, 28 days

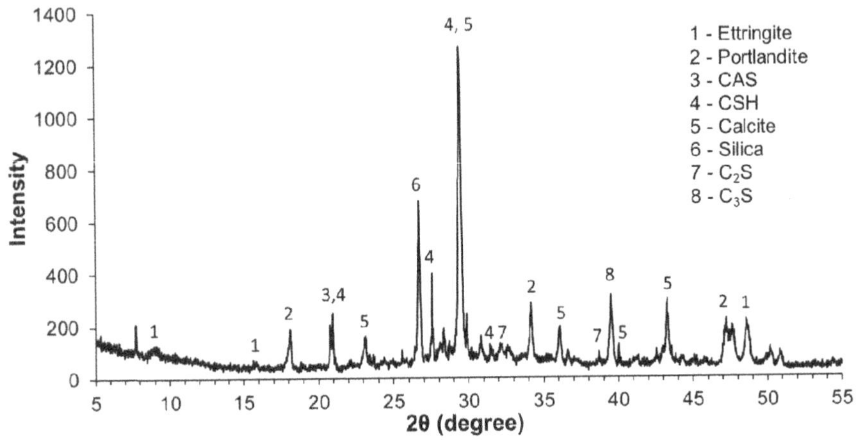

Fig. 12 XRD analysis—formulation C + SR + CW, 28 days

A diffractogram at 28 days for C + SR + W and C + SR + CW samples is expressed in Figs. 11 and 12, respectively. It is possible to perceive etringite, portlandite, CAS, CSH, calcite, silica, C_2S and C_3S peaks in the figures, which were expected for products from anhydrous cement as well as from its hydration process. These results are corroborated with those presented in Fig. 13, in which the SEM of the cement paste and cassava wastewater is presented at 28 days, and it is possible to perceive the presence of etringite, portlandite and CSH, which is the main factor responsible for the mechanical strength. These phases are in accordance with what is expected in the literature regarding cement hydration products, indicating that cement reactions are occurring in the presence of cassava wastewater [48].

4.6.7 Dynamic Modulus of Elasticity

Figure 14a shows the dynamic elasticity module of the C + SR + CW, C + SR + W, C + S + CW and C + S + W formulations. When analyzing the results, it is noticed that the sample with the presence of SR and cassava wastewater showed a higher elasticity module, agreeing with the higher density in the hardened state of these samples and greater compression strength. When analyzing the presence of SR separately (C + SR + W, C + S + W), it is possible to perceive the tendency of greater modulus in the sample with the SR being related to smaller voids and granulometric adjustment. In observing the isolated presence of the cassava wastewater (C + S + W, C + S + CW), the greatest elasticity modulus of the samples with the presence of the cassava wastewater can be observed. This behavior is indicative of the filler effect of the cassava particles present in the cassava wastewater occupying voids, thus enabling greater wave speed to propagate in the mortars.

In addition, the presence of more intense CSH peaks indicates its greater hydration and consequently greater density, which in turn contributes to propagating waves.

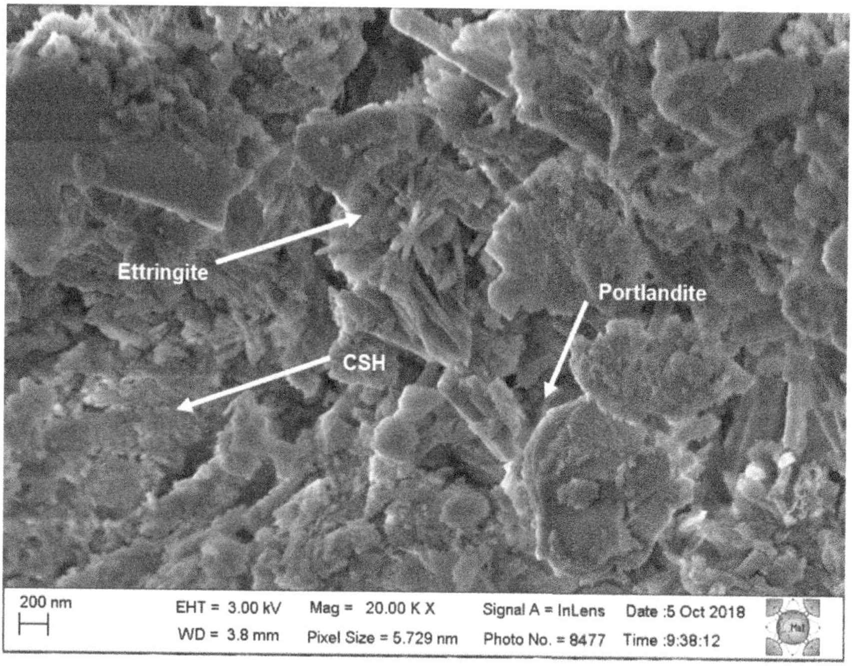

Fig. 13 SEM observation of cement and cassava wastewater mixture 28 days

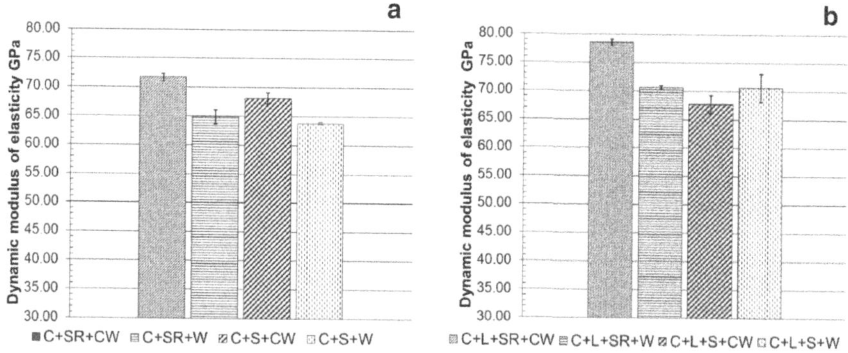

Note: C — cement; L- lime; S — Natural Sand; SR — Scheelite residue 75% gross and 25% fine; W — water; CW - Cassava wastewater.

Fig. 14 Dynamic modulus of elasticity for the proportions 1:3 (**a**) and 1:1:6 (**b**)

These results contribute to the compressive strength results, in which the samples with SR and/or cassava wastewater showed greater strength and consequent tendency to greater modulus of elasticity. Thus, the substitution of fine aggregate by SR or hydration water by cassava wastewater did not compromise the elasticity modulus of the mortars.

The dynamic elasticity module for the C + L + SR + CW, C + L + SR + W, C + L + S + CW and C + L + S + W formulations is shown in Fig. 14b. When analyzing the presence of SR separately (C + L + SR + W, C + L + S + W), equivalent values of the modulus of elasticity (70.62 ± 0.29 GPa) are presented for C + L + SR + W and (70.62 ± 2.46 GPa) for C + L + S + W. The isolated presence of cassava wastewater in the formulations (C + L + S + CW, C + L + S + W) presented equivalent elastic modulus (67.69 ± 1.54 GPa) for C + L + S + CW and 70.62 ± 2.46 GPa for C + L + S + W. The biggest difference occurred in the joint presence of SR and cassava wastewater in the C + L + SR + CW (71.67 ± 0.62 GPa) and C + L + SR + W (64.88 ± 1.20 GPa) formulations. In a similar way to cement-only formulations, the results for the elasticity modulus of 1:1:6 proportions presented results which follow the tendency to increase and decrease which occurred in the compressive strength.

The results also showed a similar behavior to those presented by Medeiros [7] when replacing 100% of the natural aggregate with the SR, as well as the modulus of elasticity values between the 1:3 and 1:1:6 mixtures. The results are similar to the density tests (Fig. 7) and voids index (Fig. 8), in which there is little variation in density and voids between the analyzed formulations; such factors are significant for the elasticity module [70]. In addition, the modulus of elasticity is predominantly influenced by the transition zone [71].

Recena [38] comments on the relationship between the modulus of elasticity and compression strength, but highlights how sensitive this result is, and how often this correlation ends up being difficult.

Regarding water interference, Kadir [72] analyzed concretes with partial replacement of hydration water (up to 50%) with washing water from a concrete mixer truck. The results showed a tendency to obtain a lower elasticity module with a greater participation of the waste used.

5 Conclusions

A predominance of calcium oxide, silicon dioxide, iron oxide and aluminum oxide was found in the composition of the scheelite residue through the analysis by X-ray fluorescence (XRF) and X-ray diffraction (XRD), while cassava wastewater showed higher nitrogen and calcium concentrations.

Granulometric adjustment with 75% of the coarse fraction of the coarse scheelite residue and 25% of the coarse scheelite residue enabled better particle packing and a greater specific gravity (3.04 g/cm^3).

Cassava wastewater did not compromise the cement hydration. In addition, the cassava wastewater showed modifying behavior of the mortar plasticity with less liquid/dry material ratio, contributing to incorporating air into mortars in the fresh state. The total solids present in the cassava wastewater, as well as the starch, showed indications of contributing to the mortar strength and absorption.

All mortars with SR showed higher mass densities in the hardened state when compared to those produced with natural sand, with the highest density being the C + SR + W composition with 2112.25 ± 19.30 kg/m^3. The formulations using cassava wastewater as hydration water showed lower densities in the hardened state.

The absorption by immersion was reduced in the 1:3 compositions with the presence of cassava wastewater and SR. Capillary absorption was significantly reduced with the presence of cassava wastewater and SR. Mortar voids were reduced with the presence of the SR provided by the performed granulometric adjustment.

The compressive and tensile strength of mortars showed greater increases when used together with SR and cassava wastewater. However, mortars which used SR or cassava wastewater in isolation also showed greater strength. The highest compressive strength obtained occurred for the C + SR + CW composition with 17.40 ± 0.31 MPa, while the C + SR + CW composition had the highest tensile strength with 4.90 ± 0.16 MPa, both at 28 days.

The presence of cassava wastewater in the compositions enabled higher CSH peaks, which are mainly responsible for cement strength, in addition to not compromising the formation of hydrated cement products, CSH, etringite and portlandite, as indicated by SEM. The calcite peaks ($CaCO_3$) also showed higher intensities associated with the cement hydration product and the presence of SR.

The dynamic elastic modulus in the 1:3 formulations showed higher values for samples with the presence of cassava wastewater, with the highest value obtained for the formulation with SR and cassava wastewater (C + SR + CW). The greatest modulus of elasticity also occurred for the formulation with SR and cassava wastewater (C + L + SR + CW) in 1:1:6 compositions.

In summary, the replacement of natural aggregate by SR and hydration water by cassava wastewater was important for improving the physical and mechanical properties of mortars in the fresh and hardened state. SR played a contributory role in developing a denser and stronger mortar. The presence of cassava wastewater led to incorporating air, thus reducing density, but at the same time, this increased the strength of mortars. Thus, the joint presence of the SR and cassava wastewater enables producing a lighter mortar with greater strength, making it an innovative and attractive material for the construction industry. An invention patent application was filed with the National Institute of Intellectual Property in Brazil (INPI) registered under number BR 10 2019 019978 4.

Acknowledgements This study was financed in part by the *Coordenacao de Aperfeicoamento de Pessoal de Nivel Superior*—Brazil (CAPES)—Finance code 001, the Federal Institute of Education, Science and Technology of Paraiba and the Federal University of Rio Grande do Norte.

References

1. Carasek HM (2010) At ISAIA In: Geraldo (ed) Civil construction materials and principles of materials science and engineering (Chapter 26), 2edn. IBRACON, vol 1 and 2, pp 863–891

2. Anepac (2015) The aggregate market in Brazil. Available: http://www.anepac.org.br/agregados/mercado/item/download/69_04062b071b7171f3481b7a0e8f36f5ac. Accessed 30 Oct 2018
3. Braga AM, Silvestre JD, De Brito J (2017) Compared environmental and economic impact from cradle to gate of concrete with natural and recycled coarse aggregates. J Cleaner Prod 162:529–543
4. Brazil (2018) National Department of Mineral Production. Mineral Summary 2016. DNPM.131 p
5. Brazil (2017) National Department of Mineral Production (DNPM) Mineral Yearbook of Rio Grande do Norte. 20 p
6. Cano TM, Costa JL, Nesi JR (2009) Tungsten. Mineral economy of Brazil. National Department of Mineral Production, Brasília, pp 148–167
7. Medeiros M (2016) Studies of coating mortars with scheelite residue. F. Dissertation (Master in Civil Engineering). Federal University of Rio Grande do Norte. Technology Center. Graduate Program in Civil Engineering. Natal/RN
8. Pereira BRF (2019) Scheelite mining in the Semi-arid Potiguar: evaluation of quality and environmental perception in areas degraded by this anthropic activity. (Master). Federal University of Rio Grande do Norte. Biosciences Center. Regional Graduate Program in Development and Environment. Natal/RN
9. Gerab AT (2014) Use of the large residue from scheelite processing in road applications. Masters dissertation. Federal University of Rio Grande do Norte. Technology Center. Graduate Program in Civil Engineering. Natal/RN, 106 p
10. Paiva EHG (2013) Evaluation of Portland cement concrete with scheelite production residue to replace fine aggregate. Masters dissertation. Federal University of Rio Grande do Norte. Technology Center. Graduate Program in Civil Engineering. Natal/RN. 106 p
11. Silva LEB et al (2018) General aspects and peculiarities about cassava (Manihot esculenta Crantz). Diversitas J 3(1):13–23
12. Madeira JGF et al (2017) Ecological analysis of hydrogen production via biogas steam reforming from cassava flour processing wastewater. J Cleaner Prod 162:709–716
13. Silva FFD, Bertonha A, Freitas PSL, Muniz AS, Rezende R, Gonçalves ACA, …and Neto VSC (2005) Splitting a starch into glucose to identify residential water from the cassava industry in the soil profile. Acta Sci Agron Maringá 27(3):507–511
14. Peres S et al., de Los Angeles Perez Fernandez Palha M (2019) Anaerobic digestion process for the production of biogas from cassava and sewage treatment plant sludge in Brazil. Bioenergy Res 12:150–157
15. FAOSTAT (2018) Food and agriculture organization of the United Nations (FAOSTAT) statistics database. FAO, Rome, Italy. Available at: http://www.fao.org/faostat/en/#data. Accessed 23 Jan 2019
16. De Carvalho JD et al (2017) Biological aspects and insecticide action of plant species on eggs and nymphs of citrus black fly (Aleurocanthus Woglumi Ashby-Aleyrodidae) at laboratory level. Braz Mag Fruticulture 39(5)
17. Ribas MMF, Cereda MP, Villas Bôas RL (2010) Use of cassava wastewater treated anaerobically with alkaline agents as fertilizer for maize (Zea mays L.). Braz Arch Biol Technol 53(1):55–62
18. Okudoh Vincent et al (2014) The potential of cassava biomass and applicable technologies for sustainable biogas production in South Africa: a review. Renew Sustain Energy Rev 39:1035–1052
19. de Chaves YAO et al (2019) Performance study of a microturbine system for cogeneration application using biogas from manipueira. BioEnergy Res, 1–9
20. Mombo S et al (2017) A socio-scientific analysis of the environmental and health benefits as well as potential risks of cassava production and consumption. Environ Sci Pollut Res 24(6):5207–5221
21. Andreani Cristiane L et al (2019) Impact of operational conditions on development of the hydrogen-producing microbial consortium in an AnSBBR from cassava wastewater rich in lactic acid. Int J Hydrogen Energy 44(3):1474–1482

22. Meier TRW et al (2020) Production of biohydrogen by an anaerobic digestion process using the residual glycerol from biodiesel production as additive to cassava wastewater. J Cleaner Prod, 120833
23. Wadjeam P et al (2019) Co-digestion of cassava starch wastewater with buffalo dung for bio-hydrogen production. Int J Hydrogen Energy 44(29):14694–14706
24. Zevallos DP, Querol MP, Ambrogi B (2018) Cassava wastewater as a natural pesticide: Current knowledge and challenges for broader utilisation. Annals of Applied Biology 173(3):191–201
25. Brazil (2018) National Water Agency. Environmental economic accounts for water in Brazil 2013. ANA, Brasília
26. Asadollahfardi G et al (2015) Experimental and statistical studies on the use of washing water for pre-mixed concrete trucks and a batch plant in the production of fresh concrete. Build Constr Mater 98:305–314
27. Malaguti VS (2016) Reuse of water and waste generated by washing concrete mixers: analysis of the effect on compressive strength of machined concrete. 124 f. Master dissertation. Federal Technological University of Paraná. Graduate Program in Civil Engineering. Curitiba/PR
28. Meena K, Luhar S (2019) Effect of wastewater on properties of concrete. J Build Eng 21:106–112
29. Ghrair AM et al (2018) Influence of grey water on physical and mechanical properties of mortar and concrete mixes. Ain Shams Eng J 9:1519–1525
30. Ismail ZZ, Al-Hashmi EA (2011) Assessing the recycling potential of industrial wastewater to replace fresh water in concrete mixes: application of polyvinyl acetate resin wastewater. J Clean Prod 19(2–3):197–203
31. Brazilian Association of Technical Standards—ABNT (2006) NM 45, Aggregates - Determination of unit mass and void volume. Mercosur Standard, Rio de Janeiro
32. Brazilian Association of Technical Standards—ABNT (2003) NBR 7175, Lime hydrates for mortars – Requirements, Rio de Janeiro
33. Breitenbach SB et al (2017) Addition of porcelain polishing residue in lime based restorations. Ceramics 63:395–401
34. Brazilian Association of Technical Standards—ABNT (2005a) NBR 13279, Mortar for laying and covering walls and ceilings - Determination of tensile strength in flexion and compression, Rio de Janeiro
35. Brazilian Association of Technical Standards—ABNT (2016) NBR 13276, Mortar for laying and wall and ceiling coverings - Preparation of the mixture and determination of the consistency index, Rio de Janeiro
36. Brazilian Association of Technical Standards—ABNT (2005b) NBR 13278, Mortar for laying and covering walls and ceilings - Determination of mass density and incorporated air content, Rio de Janeiro
37. Brazilian Association of Technical Standards—ABNT (2005c) NBR 13277, Mortar for laying and covering walls and ceilings - Determination of water retention, Rio de Janeiro
38. Recena FAP (2012) Knowing mortar, 2nd edn. EDIPUCRS, Porto Alegre
39. Brazilian Association of Technical Standards—ABNT (2005d) NBR 13280, Mortar for laying and cladding walls and ceilings - Determination of the apparent mass density in the hardened state, Rio de Janeiro
40. Brazilian Association of Technical Standards—ABNT (2009a) NBR 9778, Hardened mortar and concrete - Determination of water absorption, voids index and specific mass, Rio de Janeiro
41. Brazilian Association of Technical Standards—ABNT (2005e) NBR 15259, Mortar for laying and covering walls and ceilings - Determination of water absorption by capillarity and capillarity coefficient, Rio de Janeiro
42. Brazilian Association of Technical Standards—ABNT (2009b) NBR 15630, Mortar for laying and cladding walls and ceilings - Determination of the dynamic elastic modulus through ultrasonic wave propagation, Rio de Janeiro
43. Baali L, Naceri A, Rahmouni Z, Noui Mehidi MW (2011) Experimental study of the possibility of making a mortar with ternary sand (fine natural and artificial aggregates). Physics 22:275–285

44. Brazilian Association of Technical Standards—ABNT (2017) NBR 16605, Portland cement and other powder materials - Determination of specific mass, Rio de Janeiro
45. Neves OSC, Souza AS, Costa MA, de Almeida Sousa L, Viana AES, Neves VBF (2014) Persistence of cyanide and pH stabilization in cassava wastewater. Braz J Agroindustrial Technol 8:1274–1284
46. Nasu EGCC, Formentini HM, Furlanetto C (2015) Effect of cassava wastewater on tomato plants infected by the nematode Meloidogyne incognita. Crop Prot 78:193–197
47. Brazil (1978) Ministry of Labour. NR-15-unhealthy activities and operations - annex 11
48. Mehta PK, Monteiro PJM (2014) Concrete: structure, properties and materials, 2nd edn. IBRACON, São Paulo
49. Neto MLQ et al (2016) Use of scheelite mining residues in laying and lining mortars. In: VII Brazilian Congress of Environmental Management (ConGea), 2016, Campina Grande/PB. Anais IBEAS - Brazilian Institute of Environmental Studies, Campina Grande/PB
50. Rebouças Cyrus Santos et al (2015) Use of a system to reduce the hydrocyanic acid present in the cassava wastewater using solar energy. Braz J Agroindustrial Prod Campina Grande 17(4):391–397
51. Romano, RCO (2013) Incorporation of air in cementitious materials applied in civil construction. Thesis (Doctorate). EPUSP Civil Construction Engineering Department, University of São Paulo, São Paulo, 227 p
52. Biazzon JC et al (2019) Mechanical resistance to adhesion on wooden surfaces applied and joined by PVAc adhesives. Matéria (Rio de Janeiro) 24(1):e-12323
53. Gonçalves FG et al (2016) Evaluation of the shear strength of Pinus sp. glued at room temperature. Braz J Wood Sci 7(1):42–50
54. Henrique CM, Cereda MP, Sarmento SBS (2008) Physical characteristics of biodegradable films produced from modified cassava starches. Food Sci Technol 28(1):231–240. (Brazilian Society of Food Science and Technology, Campinas, Brazil, enero-marzo)
55. Izaguirre A, Lanas J, Álvarez JI (2010) Behavior of a starch as a viscosity modifier for aerial lime-based mortars. Carbohyd Polym 80(1):222–228
56. de Souza JM (2019) Soil-cement bricks produced with cassava wastewater instead of water. Master dissertation. Federal University of Rio Grande do Norte, Technology Center, Graduate Program in Civil Engineering, Natal/RN, 80 p
57. Choi SY, Choi YS, Yang EI (2017) Effects of heavy weight waste glass recycled as fine aggregate on the mechanical properties of mortar specimens. Ann Nucl Energy 99:372–382
58. Santamaría-Vicario A, Rodríguez C, Junco S, Gutiérrez-González V Calderón (2016) Durability behavior of steelmaking slag masonry mortars. Mater Des 97:307–315
59. Carasek H, Araujo RC, Cascudo O et al (2016) Sand parameters that influence the consistency and mass density of the coating mortars. Revista Matéria 21(3):714–732
60. Eires R, Camões A, Jalali said (2010) Optimization of construction performance on terra with resource to bio-polymers. University of Minho, Department of Civil Engineering, Campus de Azurém, 4800-058, Portugal
61. Abraham SM, Ransinchung GDRN (2018) Resistance and permeation characteristics of cement mortar with aggregates recovered from asphalt pavements. Build Constr Mater 167:700–706
62. Akindahunsi KA (2019) Investigation on the use of starch extracted from cassava and corn as a mixture in the concrete creep. Build Constr Mater 214:659–667
63. Fontes WC, Mendes JC, Da Silva SN, Peixoto RAF (2016) Mortars for laying and coating produced with iron ore tailings from tailing dams. Constr Build Mater 112:988–995
64. Bederina M, Makhloufi Z, Bounoua Z, Bouziani T, Quéneudec M (2013) Effect of partial and total replacement of the river's siliceous sand with crushed limestone on the durability of mortars exposed to chemical solutions. Build Constr Mater 47:146–158
65. Ledesma EF, Jiménez JR, Fernández JM, Galvín AP, Agrela F, Barbudo A (2014) Properties of masonry mortars manufactured with fine recycled concrete aggregates. Constr Build Mater 71:289–298
66. Bianche Juliana Jerásio et al (2016) Cutting strength in the eucalyptus sp and pinus sp. Wood Revista Árvore 40(6):1109–1117

67. Andrade JJO, Wenzel MC, Rocha GH, Silva SR (2018) Performance of mortars containing sludge from water treatment plants as fine recycled aggregate. J Clean Prod 192:159–168
68. Guelmine L, Hadjab H, Benazzouk A (2016) Effect of elevated, temperatures on physical and mechanical properties of recycled rubber mortar. Constr Build Mater 126:77–85
69. Barreto MTL et al (2013) Chemical attributes of two soils applied to the handling application. Revista Brasileira de Ciências Agrárias - Brazilian Journal of Agricultural Sciences, [s.l.] 8(4):528–534
70. Farinha C, Brito J, Veiga R (2016) Incorporation of aggregates of fine sanitary wares in coating mortars. Constr Build Mater 83:194–206
71. Khyaliya RK, Kabeer KSA, Ashok Kumar VYAS (2017) Evaluation of strength and durability of lean mortar mixes containing marble waste. Constr Build Mater 147:598–607
72. Kadir AA et al (2016) The effect on slurry water as a fresh water replacement in concrete properties. In: IOP conference series: materials science and engineering. IOP Publishing, pp 012–041

Earthbag Construction System with Cassava Wastewater

Ana Ligia Pessoa Sampaio

Abstract This chapter presents a proposal for making blocks using earthbags with the incorporation of cassava wastewater. Its viability was verified by physical and mechanical tests and by making prototypes that test the system compared to conventional masonry through thermal monitoring using thermocouples. The mechanical properties obtained for the blocks with soil and cassava wastewater are comparable to its alternative with water and cement, due to the improvement in the compaction allowed by cassava wastewater and its cementing performance. Thermo-acoustic performance is also superior, proven by calculation and monitoring tests.

Keywords Earth construction · Soil · Mechanical properties · Thermal properties · Acoustic properties

1 Introduction

Intending to make a general overview of the potentialities of the use of cassava wastewater in substitution of water in civil construction, this book sought to bring possibilities that include the soil-cement-cassava wastewater, cement-cassava wastewater, cement-residue-cassava wastewater systems and, in this chapter, only soil-cassava wastewater, as an alternative to reduce or even avoid the use of cement in civil construction. Highly industrialized materials such as cement, steel, and ceramic blocks need high-energy demand to be produced in addition to the emission of carbon dioxide into the atmosphere and the release of toxins in nature. Natural resources are rapidly being consumed due to the use of large volumes of these materials, generating waste throughout the life cycle of the building, according to the Brazilian Council for Sustainable Construction [15]. Reducing the use of these inputs should be one of the main goals of the civil construction sector, not least because the scarcity of resources due to unbridled use would be detrimental to the sector itself.

A. L. P. Sampaio (✉)
AoA Arquitetura, Natal, Rio Grande Do Norte, Brazil
e-mail: analigiapsampaio@ufrn.edu.br

W. Acchar and V. Monteiro da Silva (eds.), *Use of Cassava Wastewater and Scheelite Residues in Ceramic Formulations*, Green Energy and Technology,
https://doi.org/10.1007/978-3-030-58782-6_5

In this scenario, with the resumption of ecological awareness in recent years due to the environmental and energy crisis that the world has been suffering, the discussion around the benefits of some traditional construction techniques, such as earth-based buildings, which were left aside for the sake of modernity and industrialization of the means of production, is surfacing. Onshore constructions are a set of techniques whose main raw material is soil, a material of great abundance, found worldwide, cheap, incombustible, recyclable, and reusable with effective thermo-acoustic properties [24]. Some of these techniques are as old as the notion of civilization itself. There are several examples of constructions on land that have been maintained over the millennia, and some of these works are still in use today, going against what was agreed on the fragility and little durability of this type of construction. Although always associated with a past of precarious conditions, these techniques must be modernized to the needs of the contemporary world, adapting to the requirements not only functional and structural but also esthetic, thus being able to compete with the modern techniques.

Thus, as previously discussed, the construction system proposed in this chapter seeks to value intrinsic and particular elements of each region, such as climate, territory, and culture as opposed to the mass production of buildings that is currently observed. For this reason, due to the universe of study of this paper, Natal, located in the state of Rio Grande do Norte (RN), Brazil, is part of the Bioclimatic Zone 08, and the specificities of the presented block, such as thickness and composition, were made to correspond to the recommendations of that zone according to the Brazilian Standard (NBR) 15,220 proposed by the Brazilian Association of Technical Standards [10]. As one of the advantages of using an abundant material, such as soil, in this work, we tried to bring a system with high thermal mass, thinking not only about its mechanical properties, but also valuing what is known to be some of the greatest potentials of onshore constructions: their thermal and acoustic properties.

2 Earth Construction Techniques

The earth construction techniques grew according to the cultures that developed them. However, the growth was interrupted when cement and steel techniques became popular [25], and those techniques ended up being left aside in the technological race. Thus, as highlighted by Minke [21, p. 16], "soil as a construction material has lost its credibility due to the lack of knowledge of its wide possibilities." Building with earth has become a symbol of poverty, due to the emergence of materials considered more modern in civil construction.

Brazil's most traditional earth construction techniques are wattle and daub, rammed earth and adobe, brought by the Portuguese. The wattle and daub technique is widespread in the northeast region of the country, and it is a very versatile, adaptable, and low-cost way of building houses. It is so-called due to the way it is performed, in which its structure is made with natural wooden logs found in the region, and it is filled manually with wet clay soil. The use of water with no control and the lack of strategies to reduce the shrinkage of the finished wall, which causes cracks, a solvable problem, culminated in the stigmatization of the system in the country. The rammed earth technique has a unique esthetic and does not require any type of finishing, being carried out through the application of soil in wooden forms and compacted in successive layers. Finally, the adobe technique consists of producing blocks by extrusion or compression with moist clay soil, usually with the addition of fibers [2].

Added to that is the precarious technical knowledge about the systems and the inappropriate use of that knowledge, leading to a misunderstanding of these technologies. In this vicious cycle, wattle and daub constructions became a public health problem in many areas of Brazil, not because of its performance, but due to the poorly discussed execution issues that allowed the proliferation of insects inside of its cracked walls such as the barber, the vector of the Chagas disease [19]. It reinforces its stigma of being an underdeveloped and dangerous way of building.

On the other hand, sustainable thinking emerges in a more organized way in the 1980s, with a growing search for alternatives to traditional techniques, such as those that use soil and even encouraging the emergence of specific norms [25]. This search becomes more urgent when people are aware of data such as those brought by CBCS [16] that civil construction alone is responsible for the consumption of one-third of the planet's natural resources, for 40% of energy consumption, 40% of solid waste, and consumes 12% of drinking water.

Thereby, new techniques are emerging based on the traditional ones. Earthbag constructions are structures built with bags filled with stabilized or not stabilized soil, constituting a range of bearing and non-bearing construction systems as foundations, walls, and roofs [25]. The most popular earthbag techniques are Superadobe, Hiperadobe, Brickeradobe, and bagged rammed earth.

These relatively new techniques lack deep study, making its use difficult and increasing the variability of parameters that need to be carefully analyzed, especially when considering that each soil has particularities that would require tests to know. Among its advantages, there is a low environmental impact, the possibility of execution of supporting structures in more organic formats, the solidity of the structure, and considered to be earthquake-proof [26], in addition to the low technological complexity. In Brazil, these techniques have been widespread by non-governmental organizations (NGOs) and permaculture groups [25] with other low embodied energy materials, but not as known by the population, due to low acceptance of its usually artisanal production and bias.

Despite this, the most quickly spread earthbag construction technique throughout the world was the Superadobe. Proposed by the Iranian Nader Khalil, the system consists of long sandbags filled with soil compressed in successive layers with barbed wire between them as part of a work published in a symposium entitled "Lunar bases and space activities of the twenty-first century" organized by the National Aeronautics and Space Administration (NASA) in the 1980s [20]. The aim of this technique was to find a viable way of building on the Moon and in Mars with low cost. Among other projects, he worked on the creation of accommodation in refugee camps as a consultant to NASA and the United Nations (UN).

Proposed as an evolution of the Superadobe, the Hiperadobe technique was developed by Fernando Pacheco, a Brazilian civil engineer. Pacheco changed the usual sandbag for long bags made of high-intensity polyethylene plastic fabric, called raschel, which improves the adhesion of the soil between the layers and between the wall and the plastering. Also, it is lighter and costs less than the sandbag.

Finally, the Brickeradobe technique uses the same materials as Hiperadobe and was also proposed by Fernando Pacheco with some differences. The main ones are the bag and the compression method. The technique consists of small bags filled with soil compressed using a hand press and pestle to make blocks, giving the possibility to build orthogonal walls and eliminating the need to compress the soil layer by layer.

Education, awareness, and teaching the correct way of how to build using these techniques could make them feasible to the population and guarantee a low-cost possibility. However, these techniques remain trapped in a stigma that goes beyond the social and reaches the scale of collective health, portraits of places where the need for housing linked to the lack of options was the reason for its applications. Besides, for Seco et al. [27], it is still necessary to use cement to guarantee some current quality standards, due to its limitations such as its permeability and susceptibility to water, requiring the designer to use strategies such as architectural protection elements or impermeable layers [24].

3 Cassava Wastewater

Cassava wastewater is a yellow liquid derived from the production of manioc flour by processing cassava roots (*Manihot esculenta Crantz*), an indigenous food that was incorporated into the life of those who came to inhabit Brazil. The plant can grow even in low fertility soils, and one of the main advantages of the cassava in comparison with other agricultural products is its usability [13].

Then, aiming to make earth constructions technically and functionally more feasible to use nowadays, the cassava wastewater comes to solve some of the historical problems of using clay: the cracks by retraction due to using large amounts of water to obtain more plasticity in the material and the crumbling effect when dried. In addition, civil construction consumes about 21% of all treated water on the planet, 13.6% of that due to the construction of buildings, according to the US Green Building Council. For cities that suffer from water scarcity, thinking in new ways of constructing is becoming increasingly urgent.

4 Methodology

Once defined the materials and known its limitations, the experimental stage of this work was divided into two phases, one in the laboratory and the other one testing the construction system through prototypes in loco. The laboratory phase consisted of the processing and characterization of the materials, the definition of the compositions, and the ideal moisture content of water and cassava wastewater, as well as some the technological tests. The in loco phase consisted of building physical prototypes using the constructive system proposed here to assess its thermo-acoustic performance.

4.1 First Phase

4.1.1 Description and Treatment of the Materials

The materials used were: a predominantly lateritic clay soil collected on a site located at RN-313 road, in Pium, state of Rio Grande do Norte (RN), Brazil; Portland Cement Type V, known as High Early Strength Cement (CP-V ARI MAX) from Cimento Brennands; wrap bags developed in raschel mesh with high-density polyethylene; cassava wastewater collected in a flour mill in the city of Brejinho, RN, known for the industries related to the agricultural sector; and the water used was that distributed by the local supply system, the Water and Sewerage Company of Rio Grande do Norte (CAERN).

The raschel bags come in different sizes, colors and are generally found at popular fairs for transporting heavier fruits and vegetables. The ones used in this research were obtained at the Center of Supplies of Rio Grande do Norte (CEASA). The bags have 60 cm × 35 cm, and they were cut into bands of 25 cm × 35 cm for enveloping the blocks. The remaining 10 cm were shredded by hand in different lengths and inserted into the mixtures used to prepare the blocks

Before use, both the soil and the cassava wastewater were treated. The collected soil went through the process of natural shade-drying, then sieved by sieve No. 4 (4.8 mm), to carry out the characterization tests. The cassava wastewater had its cyanide content reduced to values that could not offer any health risks. For this, the liquid was exposed to the sun during the day and kept in a covered environment at night, as the cyanide content drops in half in 24 h and to less than 1/4 in 48 h [31]. Also, the residue goes through a simple filtration process to eliminate any large suspended solids due to the presence of organic matter.

4.1.2 Characterization Tests

After the initial treatments, the characterization was performed. The cassava wastewater was subjected to cyanide and chemical composition tests. The chemical composition tests, which consist of determining the hydrogen potential, total alkalinity, electrical conductivity, volatile and fixed suspended solids, and color, were carried out at the Federal Institute of Education, Science and Technology of Paraíba (IFPB). The cyanide content test was executed by the Bioagri Laboratory in the city of Paulista, PE,

Brazil. Tests to determine the macros and micronutrients, chloride, sulfate, and phosphate, were performed at the laboratory of the Agricultural Research Corporation of Rio Grande do Norte (EMPARN) in the city of Parnamirim, RN, Brazil.

The soil was characterized by the following tests: granulometric composition according to the Brazilian standard NBR 7181 [5]; Atterberg Limits for defining the plasticity index, by determining the liquidity limit according to NBR 6459 [3], and plasticity limit according to NBR 7180 [4]; finally, tests for mineralogical analysis by X-ray diffraction (XRD) in a SHIMADZU XRD 7000 device and, for chemical analysis, by X-ray fluorescence (XRF) in an EDX-720 device. The X'Expert High-Score Plus software was used to identify the corresponding cards from the Inorganic Crystal Structure Database (ICSD).

According to NBR 7182 [6], the ideal water content was determined by the compaction test for the soil-water and the soil-cassava wastewater systems.

4.1.3 Blocks Preparation

The blocks have dimensions 25 cm × 25 cm × 15 cm and are semi-bagged with strips of raschel bags, functioning as a wrapper enveloping it. The blocks are produced from the manual pressing of the soil with a hand press and pestles.

The shape and role of the wrapper have been changed throughout the research. The original idea was to follow the Brickeradobe system, an earthbag construction technique. However, when compacting, the system suggests putting leftover bags into the block itself, requiring a very moist mixture. However, both actions could reduce the compression strength of the blocks through increasing voids, which was observed by Pedrosa and Xavier [22] who tested some compositions and different moisture contents for Brickeradobe blocks. The results for maximum compressive strength were 1.97 MPa for 32% humidity in 4 days, but it fell in the drying time to 1.30 MPa on the seventh day. Compositions with less moisture content did not reach 1 MPa of compressive strength and compositions with more moisture lost strength through the increasing stiffness with the loss of water content through the drying period, making them fragile to fractures.

Thus, in order to reduce the amount of water used to work the soil with its optimum moisture—value obtained through compaction tests—another approach was adopted in which the enveloping could be simpler, without significant leftovers of bag that could affect its resistance and still improve the adherence of the block during its laying and coating process.

The bags were cut into strips to be put into the form to envelop the soil as in a "wrap." As each raschel bag resulted in two 25 cm wide wrappers, fewer bags were used than in the conventional Brickeradobe system. The leftovers of each bag were shredded by hand during the execution of the blocks functioning as a fiber of different sizes to help demolding the blocks. All the compositions have threads from the raschel bag in the mixture, ensuring some residual strength when cracked.

Another problem observed was a lack of flexibility in the blocks during construction. The hand press would fix the dimensions of the blocks and, as a conventional masonry block, if different sizes are needed, the block would have to be sawn or broken later. As a way to make this process easier, at least one of the dimensions could be flexible to give some options to the constructor. The modification proposed in this research at the hand press let two of the dimensions to be flexible, while only one, length or width, is fixed. This improvement aims to facilitate the execution and reduce intentional cracks since the block can be produced smaller if necessary, as shown in Fig. 1

Besides, two iron pestles of different dimensions were tested. At the tests, the larger pestle required more effort to lift it due to its self-weight and transmitted less compression energy, due to the air trapped inside the walls of the hand press. To reduce the air resistance, the pestle and the walls of the hand press were pierced. Still, part of the energy was lost in the process, causing irregularities in the surface of the block. The smaller pestle had better performance and no air resistance; therefore, it was used for all blocks. Figures 2 and 3 show a scheme of the hand press and the pestles.

Therefore, the production of the blocks follows the order: delimitation of the size of the block to adjust the hand press; manufacture of the bags, on-site or in advance; adjust the bag in the press to receive the soil; fill half the height with soil and compact with 20 strokes with the pestle on each side; fill again with soil to the wanted height;

Fig. 1 Blocks with different dimensions according to the need

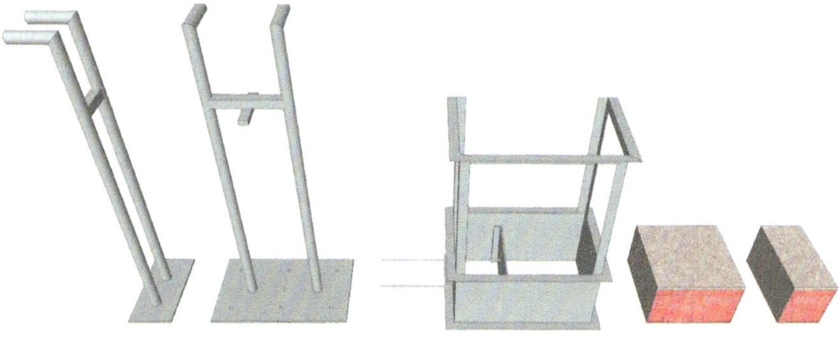

Fig. 2 Scheme of the hand press and the blocks

Fig. 3 Movable and removable wall detail

close the bag and place another 3 cm of soil; compact the excess to achieve the wanted height; and, finally, demold the block with the help of the pestle, pushing it down while the form is pulled up.

When ready, wait between 12 and 24 h to use the blocks, depending on if it is sun-drying or shade-drying. It is enough time to lose most of the surface moisture and acquire enough strength to be transported.

4.1.4 Testing the Blocks

Due to the impossibility of replicating the compression energy in specimens, normal size blocks were used for the tests. Twelve (12) blocks were molded, eight (08) for compressive strength tests, at 7 and 28 days, and four (04) for the thermal conduction and absorption test, at 28 days, for each composition. In total, 48 blocks were made for testing. The cement used for the tests was the High Early Strength Portland Cement.

The mixtures were prepared with the help of the Motomil-MB-150L concrete mixer to achieve a more uniform mix and reduce the variability of manual processes. In addition, compressive strength tests were carried out, on the dates of 7 and 28 days, using the universal press AMSLER with a 10,000 kg load cell for breaking the blocks. The test was performed according to NBR 8492 [7], except for the recommendations regarding the dimensions of the blocks.

Two caveats are important to make regarding the compressive strength test. The first is that, due to the low moisture and the plasticity of the soil, the movement of the concrete mixer generated a spherical effect on the soil particles, agglomerating them into small spheres. The second caveat to consider is that the plates responsible for applying pressure on the specimens had the dimensions of 22.5 cm $\times$ 22.5 cm, totaling 506.25 cm^2 of area, being smaller than the area of the block, which has dimensions of 25 cm $\times$ 25 cm and an area of 626 cm^2. Thus, the area for calculating the resistance was defined according to the area of the plates and not that of the blocks, as it is the effective contact area between the elements.

Tests of thermal conductivity, heat capacity, and diffusivity of the blocks were carried out to assist the evaluation of the thermal performance of the system. For that, it used the Anter Thermal Properties Corporation thermal properties meter, model Quickline^{TM}-30 accompanied by a planar sensor for ranges of thermal conductivity between 0.2 and 2 m^2 K/W, defined from preliminary tests. Its operation is based on the transient method and the temporal evolution of temperature in the distance. The test was carried out with the blocks shade-dried. However, heat capacity and diffusivity were not obtained by the meter in some blocks, mainly in the cement-cassava wastewater compositions. The process was repeated five times, and yet no value has been reached. The number of elements presented in the compositions may interfere with the meter sensors, considering that the compositions with no cassava wastewater had automatic results.

To determine the water absorption of the blocks, due to the lack of specific standards, the NBR 15,259 [11] standard was used as a reference. The norm describes the execution of the capillarity water absorption test and the capillarity coefficient for mortars through the suspension in a permeable mesh of prismatic specimens, with the water level at 5 ($\pm$1) mm above the bottom face. The specimens were cut from de-blocks and are weighed after the immersion at 10 min and at 90 min, and Eq. (1) is used to find the results:

$$\mathrm{At} = \frac{\mathrm{mt} - \mathrm{m0}}{100} \tag{1}$$

where "At" is the absorption of water by capillarity for each time (g/cm^2), "mt" is the mass of the specimen for each interval (g), and "m0" is the initial mass of the specimen (g). Obtaining the value of the difference in mass at the intervals 90 min and at 10 min, we have the capillarity coefficient.

4.2 Second Phase

Finally, for a global assessment of the system, prototypes with dimensions 1 m $\times$ 1 m indoors area and 1.5 m in height were executed. The execution process, difficulties, modifications, and thermo-acoustic performance were evaluated. In order to do that, two (02) prototypes in the form of small houses were built, one executed with the construction system proposed here, called T1, and the other in conventional masonry, called T2, for comparative purposes. For the execution of the prototype, due to its dimensions, two sizes of the blocks were chosen, a larger one of 0.25 m $\times$ 0.25 m $\times$ 0.15 m and a smaller one of 0.25 m $\times$ 0.125 m $\times$ 0.15 m for the block bonds. The houses were covered with insolated metal tiles, to reduce the impact of the roof on the temperature. Plastering and painting were applied equally to not interfere with absorptivity and heat conduction in the walls. In addition, the aim was to propose a construction system to work like any other, so it was an opportunity to observe how the system reacts in contact with plaster, paint, and adverse climatic variations.

4.2.1 Evaluation of Thermal Performance

The thermal evaluation of the system took place in two ways. The first one was mathematical and the second one through the evaluation of the prototypes.

From the data obtained through the thermal conductivity test, it was possible to numerically reach the values of thermal resistance (R) and thermal transmittance (U) of the proposed system through Eq. (2):

$$R = \mathrm{e}/\lambda \quad U = 1/\mathrm{RT} \tag{2}$$

The solar heat gain coefficient (SHGC) refers to the quotient of the solar energy absorbed by a component by the total solar energy incident on its external surface [10] and is described by Eq. (3) below, which takes into consideration the thermal transmittance of the material and the absorptivity.

$$\mathrm{SHGC} = 4 \cdot U \cdot \alpha \tag{3}$$

The thermal lag (φ) quantifies the delay for heat to flow through a constructive component. So, Eq. (4) describes thermal lag as:

$$\varphi = 1.382 \cdot \mathrm{e} \cdot \surd(\rho.c/3.6.\lambda) \tag{4}$$

The second method of evaluation was through the installation of type K thermocouples inside the prototypes. The temperature of the north and south walls were measured during five (05) days to calculate the temperature variation between them. Meanwhile, the air temperature was measured with a thermocouple placed in the center of the prototypes, at 1 m high for about ten (10) days. Eight (08) successive days were chosen to represent the week cycle. Measurements were taken every hour, and the data were stored on the Minipa datalogger thermometer, Model MT-600. The data obtained were correlated with the data emitted by the meteorological stations of the Federal University of Rio Grande do Norte (UFRN) and compared, so that it was possible to evaluate them according to the external climatic variations.

During the period from January 1st to 11th, 2020, the air temperature inside the prototypes was measured with no ventilation by blocking their entrances. From the 14th to the 26th of February, 2020, the air temperature was measured with ventilation. During this period with ventilation allowed, the temperatures of the walls were measured, from the 14th to the 20th of February, 2020, in T1 and from the 20th to the 26th of February, 2020, in T2.

4.2.2 Evaluation of the Acoustic Performance

To assess the acoustic performance, due to the proximity of a busy avenue, it was not possible to isolate the external noise or reduce its influence; thus, the values were estimated mathematically. The acoustic evaluation of the blocks took place numerically through Eq. (5), in which the frequency (f) was defined in 1000 Hz to standardize the calculation.

$$PT = 20\log(fm) - 47(dB) \tag{5}$$

It is worth mentioning that the evaluation of the performance of the prototypes did not start from the idea of isolating the parameters to evaluate the system, but from comparing the construction systems as a whole, not their isolated elements, so as not to leave the generalist proposal of this work, whose interest is to analyze all the best-known segments of civil construction assessment, as an introductory study of the proposed system.

5 Results and Discussions

5.1 *Characterization of the Soil and the Cassava Wastewater*

The studied soil has a passing percentage of 99.90% in the No. 4 sieve (4.8 mm), 69.74% in the No. 40 sieve (0.42 mm), and 27.47% in the No. 200 sieve (0.075 mm). For soil-cement compositions, NBR 10833 [8] prescribes that between 10 and 50% of the sample pass through the No. 200 sieve (0.075 mm mesh). The granulometric analysis of the soil is shown in Fig. 4.

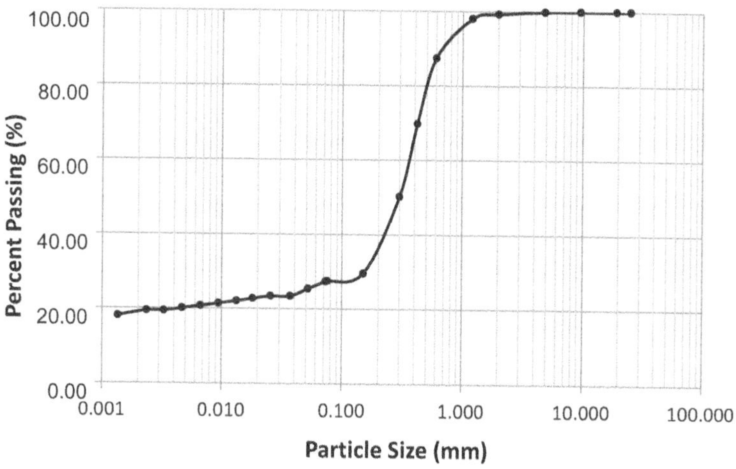

Fig. 4 Particle size distribution curve

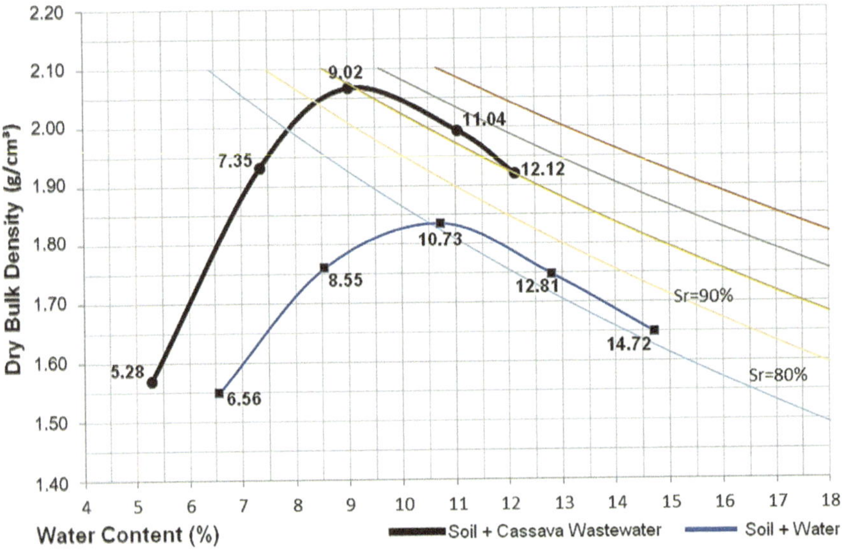

Fig. 5 Compaction test with soil, cassava wastewater, and water

This is important because the clay, although thinner than the sand, has a higher specific surface. This means that the electrostatic forces predominate in this material which explains its plasticity by the interaction between the particles and the lubricating action of the fluid that surrounds it [32].

The specific mass of the soil, determined through the pycnometer, is 2.66 g/cm^3, and the values obtained for the moisture content through the compaction test are described by the standard NBR 7182 [5], shown in Fig. 5.

For the soil-water system, the optimum moisture content obtained was 10.73%, and the dry bulk was 1.830 g/cm^3, while for the soil-cassava wastewater system the optimum moisture was 9.60% with dry bulk, 2.075 g/cm^3, an increase of 13% in comparison with the dry mass of the soil-water system with 10% less moisture. Therefore, it is observed that the soil-cassava wastewater system is denser, either through a more efficient packing or due to the possible reactions that happened in the system. Furthermore, this test helped to pin the moisture content for all the compositions at 10%, as it is an average value between the two optimum moistures and establishes a reference since it is not the focus of this work.

As for the results for the Atterberg Limits, they were obtained from tests carried out as determined by the NBR 6459 [3] and NBR 7180 [4] standards for the soil-water and soil-cassava wastewater systems. The results highlighted in Table 1 demonstrate that the values obtained followed what is specified by the NBR 10833 [8] which determines the liquidity limit and the plasticity index at values equal to or less than 45% and 18%, respectively.

The chemical analyses of the soil and the cassava wastewater were obtained using the X-ray fluorescence, and the results are given in Table 2. In it, it is possible to notice

Table 1 Atterberg limits

Atterberg limits	Soil + water (%)	Soil + cassava wastewater (%)
Liquid limit (LL)	20.6	17.3
Plastic limit (PL)	10.4	13
Plasticity index (PI)	10.4	4.3

the high concentration of iron oxide (Fe_2O_3), followed by silica (SiO_2) and alumina (Al_2O_3) in the soil, common in lateritic soils. While in the cassava wastewater, there is a predominance of calcium oxide (CaO), iron oxide (Fe_2O_3), potassium oxide (K_2O), and silica (SiO_2).

The soil was classified as lateritic, common in tropical areas because it offers favorable conditions for the intensification of chemical weathering and the presence of clay minerals, silicate minerals, hydrated metal oxides, quartz, and silica (SiO_2) in its mineralogical composition, characteristic of residual soils [17]. Lateritic soil, specifically, is formed from weathering in the upper part of the subsoil, transforming it through the laterization process [33].

It is possible to note yet that a large presence of CaO in the cassava wastewater can be related to its ease of use when it comes in contact with the ground, noticeably better when you produce blocks. In addition, Ca^{2+}, K^{+}, Na^{+}, Mg^{2+} are considered exchangeable bases, while $Al^{3+} + H^{+}$ ions are known for extractable acidity. These two values combined denote the cation-exchange capacity (CEC) of a soil and may indicate the degree of weathering, the constituent clay minerals, and, if applicable, the degree of expandability, not found in the materials used in this paper. Regarding the changes that can promote the gain of mechanical resistance of the blocks, silica and alumina minerals when present in the disorganized form in the soil in combination with the presence of calcium, an exchangeable base present in the cement, can form products known for increasing compressive strength in cementitious materials [17], either by hydration or by a pozzolanic effect, called calcium silicate hydrate (C–S–H).

At the same time, subsequent analyses were possible through the results obtained by the X-ray diffraction carried out on the soil, as can be seen in Fig. 6, which showed the predominance of peaks with higher intensities for kaolinite, as the main clay mineral ($Al_2Ca_1O_8Si_2$), quartz (SiO_2), and illite [$K(Al_4Si_2O_9(OH)_3)$]. A mineralogical analysis is important to understand which clay mineral is most prevalent in the sample, which can justify its behavior and enable us to work it more efficiently.

It is in the laterization process that the enrichment of hydrated aluminum and/or iron oxides occurs in the soil, responsible for the typical red, yellow, orange, and brown coloration, and the presence of kaolinite as the predominant clay [33]. In the natural state, they are generally unsaturated and have a low carrying capacity due to the high voids index. However, this capacity can be high when compacted, suffering a contraction in the presence of water.

Kaolinite is a clay mineral structured through 1:1 layers stacked in a regular manner that consists of a layer of $SiSO_4$ tetrahedrons and one of $Al_2(OH)6$ octahedrons, linked by common oxygen atoms in a single layer, a strongly polar structure

Table 2 X-ray fluorescence results

Material	Fe_2O_3 (%)	(SiO_2) (%)	Al_2O_3 (%)	CaO (%)	K_2O (%)	Na_2O (%)	ZrO_2 (%)	BrO_2 (%)	P_2O_5 (%)	CuO (%)	MgO (%)
Soil	42.8	23.6	15.1	2.16	5.34	–	4.20	–	–	–	0.71
Cassava W.	27.5	8.18	1.80	29.4	17.2	5.82	–	3.20	2.45	1.94	–

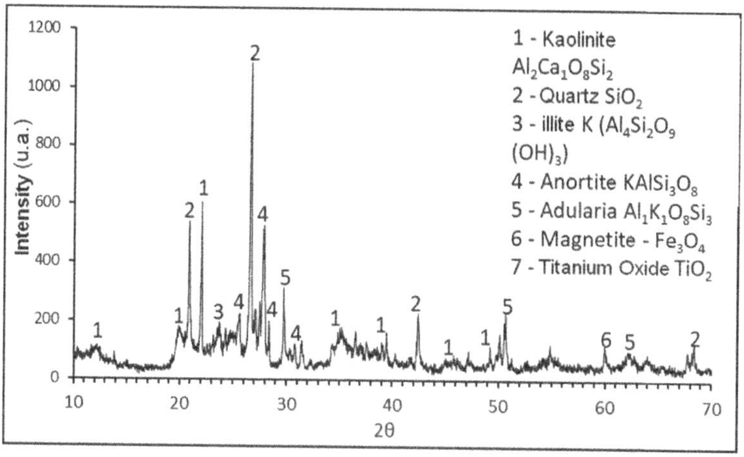

Fig. 6 X-ray diffraction of the soil

[32]. This group of clay minerals combined with oxides and hydroxides of iron and/or aluminum in the presence of water form stable aggregations that promote the natural cementing of the particles due to the hydration process of the oxides and hydroxides which, when covering the clay minerals, reduce their adsorption capacity of water [33].

This cementation process is aimed in this study with the perspective that the acidic medium present in the cassava wastewater would function as an accelerator of chemical reactions between existing clay minerals in the soil used. That could favor cation exchanges and support other reactions to happen that could increase its compressive strength. The chemical composition of the cassava wastewater used in the research is described in Table 3.

The chemical elements given in Table 3 were obtained by laboratory tests and compared to two references [18, 23]. The acidity of the cassava wastewater is noticeable along with the high concentrations of potassium, nitrogen, sodium, iron, and calcium. The concentration of cyanide was according to the recommendation, whose maximum concentration limits established by both Occupational Safety and Health Administration (OSHA) and by Regulatory Standard NR 15 [1] are 8 parts per million (ppm) and 10 ppm, respectively, of exposure for a workload of 8 h/day. The values for COD and BOD were not obtained through laboratory tests because they were not part of the scope of this work, but it is given in Table 3 for reference.

5.1.1 Compositions

Based on the characterization tests and determination of the optimum moisture content, four (04) formulations were defined and listed in Table 4.

Table 3 Chemical composition of cassava wastewater

Chemical elements	Silva et al. [28]	Peres et al. [23]	This work (2020)
Nitrogen (mg/L)	32.40	–	1121
Phosphorus (mg/L)	17.80	–	132
Potassium (mg/L)	333.60	–	1456
Calcium (mg/L)	31.37	–	93
Magnesium (mg/L)	36.87	–	219
Sodium (mg/L)	51.70	–	351
Zinc (mg/L)	0.59	–	20
Copper (mg/L)	0.05	–	9
Iron (mg/L)	6.09	–	117
Manganese (mg/L)	0.62	–	15
Total solids (g/L)	–	92.90	14.56
Suspended solids (g/L)	–	–	2.37
Chloride (mg/L)	–	–	545.40
Sulfur (mg/L)	–	–	78.53
Phosphate (mg/L)	–	–	27.07
Cyanide (mg/L)	12	16.66	5.62
COD (g O2/L)	14.70	101.38	–
BOD (g O2/L)	6.21	29.20	–
pH	4.80	3.90	4.5

Table 4 Nomenclature of the compositions

No	Nomenclatures	Compositions
1	SW	Soil + water
2	SC	Soil + cassava wastewater
3	CW	Soil 90% + 10% cement + Water
4	CC	Soil 90% + 10% cement + cassava wastewater

Cement compositions were restricted to 10%, because due to the high mass of each block, its production would require a large amount of cement, losing its feasibility, whether financially or environmentally. However, for structural blocks that may replace other entire cement products, the evaluation of other compositions would be important, since the compressive strength gain proposed here does not come only from the hydration of the cement, but also from the compaction energy, which reduces the need for cement products to obtain the compressive strength required.

5.2 Technological Tests

5.2.1 Compressive Strength Test

All compositions submitted to the compressive strength test met the minimum value of 1 MPa established by NBR 13553 [9] for materials used in soil-cement walls without structural function, as detailed in Table 5.

It is observed that the results for compositions with cassava wastewater were superior to those with water, especially the soil-cassava wastewater composition, which had similar results to the soil-cement-water system. Then, it is noticed the increase in compressive strength with the use of cassava wastewater and cement at 28 days.

As the High Early Strength Portland Cement was used, the compressive strength at 7 and 28 days are similar, with little increase. In the CW composition, there was a drop in the compressive strength. As a hypothesis, it is possible that as the cement content is small in comparison with the soil and the spherical effect generated by the movement of the concrete mixer, it had interfered in the compaction of the blocks, during production, considering that these small agglomerated and little compacted spheres contribute to the increase of voids in the structure of the blocks. Also, cement

Table 5 Compressive strength test

Composition		Area (cm^2)	Load (kgf)		Strength (MPa)		Average strength (MPa)	
			7 days	28 days	7 days	28 days	7 days	28 days
SW	SW_1	506.25	7650	8940	1.51	1.77	1.66	1.52
	SW_2	506.25	12,460	10,240	2.46	2.02		
	SW_3	506.25	8000	6550	1.58	1.29		
	SW_4	506.25	5540	5080	1.09	1		
SC	SC_1	506.25	14,200	14,270	2.80	2.82	2.12	2.09
	SC_2	506.25	9700	7020	1.92	1.39		
	SC_3	506.25	7800	10,850	1.54	2.14		
	SC_4	506.25	11,220	10,220	2.22	2.02		
CW	CW_1	506.25	13,400	7250	2.65	1.43	3.31	2.04
	CW_2	506.25	19,930	9680	3.94	1.91		
	CW_3	506.25	15,700	9660	3.10	1.91		
	CW_4	506.25	17,910	14,650	3.54	2.89		
CC	CC_1	506.25	11,040	10,190	2.18	2.01	2.34	2.41
	CC_2	506.25	11,580	11,470	2.29	2.27		
	CC_3	506.25	11,000	9400	2.17	1.86		
	CC_4	506.25	13,820	17,720	2.73	3.50		

particles may have formed separate granular units due to their fineness, floating in the soil matrix and decreasing the system's rigidity [17].

In this sense, at 7 days, the block still had a considerable amount of water in it, leaving the clay matrix more plastic, and, despite deforming under compression, it does not break and still connects one cement unit to another that was previously floating in the system. This was observed in the test of blocks CW_2 and CW_4 at 7 days, which suffered intense deformation before the press identified the rupture of the samples.

However, at 28 days, the clay matrix would have lost most of its moisture, constituting points of weakness that left the resistance of some blocks similar to the blocks that had only soil and water. In the cassava wastewater systems, greater stability is observed between the compressive strengths at 7 and 28 days, probably due to the acidic medium that may have helped to break down the granular cement units, or even helped in their reactivity with the soil matrix.

Along with this, the presence of small particles in the cassava wastewater, in a concentration of 14,563 mg/L of total solids, may be acting as a micro filler and improving the packaging when filling the voids. Among these solids, there are also in the composition of the cassava wastewater starch and sugars that can reduce the fragility of the clay matrix by functioning as biopolymers that, even when drying, can contribute to the strength of the blocks [29], as is possible to notice in the soil-cement-cassava wastewater composition, which presented the best results for the compression test.

5.2.2 Density (Volumetric Mass)

Considering the variability aspect, due to the manufacturing process of the blocks using non-standard human strength, all blocks were measured and weighed after 28 days to observe the degree of variability between the different compositions and between samples of the same composition. Observing the results, it is possible to notice that there is higher variability between the blocks with water composition than the ones with cassava wastewater, as can be seen in Table 6, in which the mass varies more proportionally to its volume.

The variability of the systems happens due to problems in height standardization, one of the parameters that need improvements in the formwork, because, while the other dimensions are controllable, the height of the blocks is still defined without much precision by the operator.

Therefore, it is likely that this little bit higher variation in density between compositions with water is related to the ease of compaction caused by the cassava wastewater, better-distributing efforts.

Table 6 Mass/volume ratio

CP	Mass (g)	Width (cm)	Length (cm)	Height (cm)	Density (g/cm^3)	
SW_1	15,220	25	25	15.5	1.5711	1.597
SW_2	16,648	25	25	16	1.6648	
SW_3	15,328	25	25	15.5	1.5822	
SW_4	15,213	25	25	15.5	1.5704	
SC_1	17,900	25	25	15.3	1.8719	1.885
SC_2	18,500	25	25	15.5	1.9097	
SC_3	17,339	25	25	15	1.8495	
SC_4	18,497	25	25	15.5	1.9094	
CW_1	17,347	25	25	15	1.8503	1.805
CW_2	17,534	25	25	15	1.8703	
CW_3	17,401	25	25	16	1.7401	
CW_4	17,026	25	25	15.5	1.7575	
CC_1	16,281	25	25	15.2	1.7138	1.723
CC_2	16,814	25	25	15.5	1.7356	
CC_3	16,538	25	25	15.5	1.7071	
CC_4	16,268	25	25	15	1.7353	

5.2.3 Thermal Conductivity Test

The thermal conductivity (λ), the volumetric heat capacity (C), the specific heat (c), and the diffusivity (α) were the parameters obtained through the thermal conductivity test. The results are described in Table 7.

Through the results, it is noted that the mixtures with cement showed lower conductivity than compositions without it. This is due to the decrease in the permeability of the blocks without the incorporation of cement due to better compaction. The blocks only with soil, regardless of having cassava wastewater or not, are more permeable and, due to their more clayey texture, retain more water, maintaining more moisture inside, which may have increased the thermal conductivity of the block and the thermal capacity, which determines the amount of heat necessary for the temperature of a body to varying. Also, if the values obtained for two the soil-cassava wastewater blocks that the meter was able to define are observed, its higher thermal conductivity is related to its density, since the compacted soil, with less air inside, also facilitates the conduction of heat.

5.2.4 Water Absorption Test

Figure 7 shows some samples used in the capillarity absorption test, and Table 8 summarizes all the results obtained.

Table 7 Thermal properties of the blocks

CP	d (g/cm^3)	λ (W/m K)	$C\rho$ (E + 6 J/m^3 K)	c (J/Kg.K)	∝ (E-6 m^2/s)	$d$avg	λavg
SW_1	1.5711	0.725	1.44	0.946	0.503	1.597	0.701
SW_2	1.6648	0.993	1.58	0.949	0.63		
SW_3	1.5822	0.533	1.44	0.939	0.454		
SW_4	1.5704	0.553	1.41	0.927	0.392		
SC_1	1.8719	0.922	–	–	–	1.885	1.082
SC_2	1.9097	1.4	1.43	0.773	0.983		
SC_3	1.8495	0.817	–	–	–		
SC_4	1.9094	1.19	1.4	0.757	0.848		
CW_1	1.8503	0.639	–	–	–	1.805	0.672
CW_2	1.8703	0.725	–	–	–		
CW_3	1.7401	0.687	–	–	–		
CW_4	1.7575	0.636	1.39	0.816	0.459		
CC_1	1.7138	0.679	–	–	–	1.723	0.602
CC_2	1.7356	0.588	–	–	–		
CC_3	1.7071	0.616	–	–	–		
CC_4	1.7353	0.524	–	–	–		

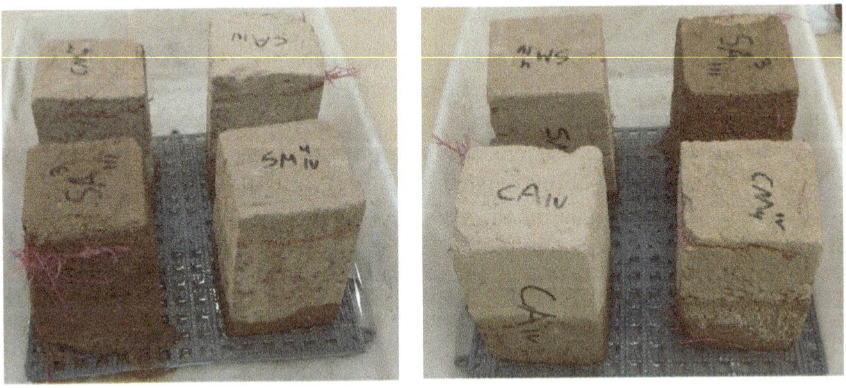

Fig. 7 End of the water absorption by capillarity test

With the results given in Table 8, it is noted the intrinsic relationship between density, even if only volumetric, and the absorption of the samples. The best results were obtained by the composition with only soil and cassava wastewater (SC), due to the better compaction and, consequently, the smaller amount of internal voids. This also suggests the binding characteristics of cassava wastewater. The cement samples obtained intermediate values between the SC and SW compositions, which had the lowest and the highest absorption rates, respectively. The samples with only soil and

Table 8 Water absorption by capillarity results

CP	d (g/cm^3)	m (g)	m_{0_seca} (g)	Moisture (%)	m_{t10} (g)	m_{t90} (g)	A_{t10} (g/cm^2)	A_{t90} (g/cm^2)	C (g/dm^2 min$^{1/2}$)	Avg. A_{t10}	Avg. A_{t90}	Avg. C
SW_1	1.5711	2139	2107	1.50	2336	2474	2.29	3.67	138	2.05	3.47	142
SW_2	1.6648	2444	2411	1.35	–	–	–	–	–			
SW_3	1.5822	2600	2574	1	2771	2944	1.97	3.7	173			
SW_4	1.5704	2327	2284	1.85	2472	2587	1.88	3.03	115			
SC_1	1.8719	2783	2752	1.11	2802	2829	0.5	0.77	27	0.43	0.67	23.75
SC_2	1.9097	2948	2926	0.75	2970	2995	0.44	0.69	25			
SC_3	1.8495	2769	2749	0.72	2798	2821	0.49	0.72	23			
SC_4	1.9094	2888	2864	0.83	2894	2914	0.3	0.5	20			
CW_1	1.8503	2637	2554	3.15	2600	2639	0.46	0.85	39	0.79	1.28	49.75
CW_2	1.8703	2556	2474	3.21	2540	2593	0.66	1.19	53			
CW_3	1.7401	2721	2619	3.75	2708	2766	0.89	1.47	58			
CW_4	1.7575	2676	2592	3.14	2705	2754	1.13	1.62	49			
CC_1	1.7138	2554	2512	1.64	2635	2684	1.23	1.72	49	1.09	1.50	41.25
CC_2	1.7356	2632	2591	1.56	2673	2714	0.82	1.23	41			
CC_3	1.7071	2393	2353	1.67	2470	2509	1.17	1.56	39			
CC_4	1.7353	2354	2298	2.38	2410	2446	1.12	1.48	36			

water were completely saturated during the test, as can be seen in Fig. 7 and suffered mass losses in their performance.

6 Thermo-acoustic Evaluation of the System

The globalization process observed around the world, combined with rapid technological advances, seems to have set aside some extremely important issues for sustainable development that are intended to solve the particular problems of each region. In this sense, assessing the performance and efficiency of buildings and their construction system is essential to propose techniques that value the local context and with the possibility of flexibility. Therefore, the soil-cassava wastewater composition was chosen to be the block formulation to build a prototype at the UFRN and be monitored using thermocouples in order to evaluate its thermo-acoustic performance.

For this evaluation, the prototypes were executed on the ground of the Federal University of Rio Grande do Norte with internal areas of 1 m^2 and 1.50 m in height, as highlighted in Fig. 8.

The prototype (Fig. 8a) called T1 was made with the blocks presented in this chapter and plastered with conventional mortar (sand, cement, and water). The prototype (Fig. 8b) called T2 was made in conventional masonry, with eight-hole bricks, and also plastered with the same mortar as T1.

6.1 Thermal Evaluation of Construction Systems

6.1.1 Thermal Resistance (R) and Thermal Transmittance (U)

The total thermal resistance (R_T), which takes into account the plaster, is obtained through Eq. (2) and by adding the resistances of homogeneous layers as given in Table 9 for each composition. For calculation purposes, it was considered 5 cm of plaster and thermal conductivity of 7 W/m K.

Fig. 8 **a** T1, earthbag system. **b** T2, masonry system. **c** T1 and T2

Table 9 Thermal resistance and thermal transmittance of the blocks

CP	d (g/cm^3)	λ (W/m K)	C (E+6 J/m^3 K)	c (J/Kg K)	$\propto$ (E-6 m^2/s)	R (m^2 k/W)	R_{total}	U [W/(m^2 K)]	λ_{avg}	U_{avg}
SW_1	1.5711	0.725	1.44	0.946	0.503	0.345	0.488	2.051	0.701	1.975
SW_2	1.6648	0.993	1.8	0.949	0.63	0.252	0.395	2.534		
SW_3	1.5822	0.533	1.44	0.939	0.454	0.469	0.612	1.634		
SW_4	1.5704	0.553	1.41	0.927	0.392	0.452	0.595	1.681		
SC_1	1.8719	0.922	–	–	–	0.271	0.414	2.415	1.082	2.647
SC_2	1.9097	1.4	1.43	0.773	0.983	0.179	0.321	3.111		
SC_3	1.8495	0.817	–	–	–	0.306	0.449	2.228		
SC_4	1.9094	1.19	1.4	0.757	0.848	0.210	0.353	2.833		
CW_1	1.8503	0.639	–	–	–	0.391	0.534	1.872	0.672	1.941
CW_2	1.8703	0.725	–	–	–	0.345	0.488	2.051		
CW_3	1.7401	0.687	–	–	–	0.364	0.507	1.973		
CW_4	1.7575	0.636	1.39	0.816	0.459	0.393	0.536	1.866		
CC_1	1.7138	0.679	–	–	–	0.368	0.511	1.957	0.602	1.788
CC_2	1.7356	0.588	–	–	–	0.425	0.568	1.760		
CC_3	1.7071	0.616	–	–	–	0.406	0.549	1.822		
CC_4	1.7353	0.524	–	–	–	0.477	0.620	1.613		

All average values obtained for thermal transmittance (U) are according to the requirements of the Bioclimatic Zone 08, where the prototypes were built, which values must be below 3.60 W/(m^2 K) according to NBR 15220 [10]. As Natal, RN, Brazil, is a humid city, the conductivity of the blocks becomes higher than if the experiment happened in a dry climate city, as the cities located at the Bioclimatic Zone 07, which happens because the particles of water in the air are absorbed by the blocks, keeping them always moisturized and increasing its thermal conductivity when filling the pores that previously had only air, relation confirmed by Zimmer [34] in their studies.

In summary, the thermal conductivity values presented above would be different in Bioclimatic Zone 07 due to the difference in air humidity. That is why, it is so important to evaluate the performance of each construction technique according to the local conditions. The capacity of the block of absorbing water and its large volume could cool the system through evaporation, an endothermic reaction that captures the heat of the environment, cooling it in an effect commonly known as "evaporative cold."

6.1.2 Solar Heat Gain Coefficient (SHGC)

As the walls were plastered and painted white, it was defined the absorptance value as the one obtained by Castro et al. [14] through spectrophotometric analysis of 0.25 for white acrylic paint, and the results obtained with the help of Eq. (3) are highlighted in Table 10.

The standard NBR 15220 [10] establishes that the solar heat gain coefficient must be equal to or less than 3.50 for Bioclimatic Zone 07 and equal to or less than 4 for Bioclimatic Zone 08. The low absorptance values were expected due to the white acrylic color. Then, all compositions reached the recommended values. For more enhanced effects, darker colors would be ideal.

6.1.3 Thermal Lag (φ)

As Eq. (4) for thermal lag requires values of the thermal capacity or the specific heat of the materials, the calculations were made only for those compositions that the thermal meter was able to quantify it. Table 11 shows the results.

Due to the high thermal mass, all compositions had thermal lags higher than the minimum established by the NBR 15220 standard [10]. Thick walls, like those studied in this chapter, promote thermal inertia, a parameter that involves two main phenomena, the thermal lag that leads the heat gains and losses throughout the day, and the reduction in thermal amplitude, exactly for balancing the external temperature variation.

This happens because, when the external temperature is higher than the internal one, the heat flows from outside to inside. On a thin wall, this process happens

Table 10 Solar heat gain coefficient (SHGC)

CP	U [W/(m^2 K)]	SHGC	U_{avg}	SHGC$_{avg}$
SW_1	2.051	1.640	1.975	1.580
SW_2	2.534	2.027		
SW_3	1.634	1.307		
SW_4	1681	1.345		
SC_1	2.415	1.932	2.647	2.118
SC_2	3.111	2.489		
SC_3	2.228	1.782		
SC_4	2.833	2.267		
CW_1	1.872	1.498	1.941	1.552
CW_2	2.051	1.640		
CW_3	1.973	1.579		
CW_4	1.866	1.493		
CC_1	1.957	1.565	1.788	1.31
CC_2	1.760	1.408		
CC_3	1.822	1.458		
CC_4	1.613	1.290		

Table 11 Thermal lag of the blocks

CP	d (kg/m^3)	λ (W/m K)	c (J/Kg K)	φ (h)
SW_1	1571.097	0.725	0.946	8.245
SW_2	1664.800	0.993	0.949	7.264
SW_3	1582.245	0.533	0.939	9.616
SW_4	1570.374	0.553	0.927	9.342
SC_2	1909.677	1.400	0.773	5.913
SC_4	1909.368	1.190	0.757	6.346
CW_4	1757.523	0.636	0.816	8.649

quickly, but on a thicker wall, it can last long enough until the heat starts to flow back before reaching inside completely, at night, for example.

6.1.4 Thermal Monitoring of Prototypes

In order to establish a reference when creating the graphs, the period was defined in eight days, always starting and ending at 0:00. Figure 9 shows the results of monitoring without ventilation for T1 and T2, from 3rd to 10th of January of 2020. The T1 values are related to the air temperatures of the prototype executed with

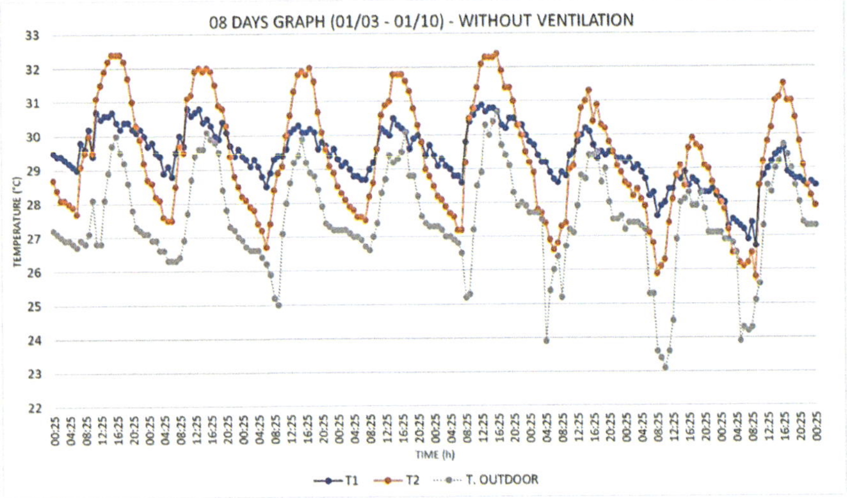

Fig. 9 Air temperature monitoring without ventilation

the construction system proposed in this research, and the T2 values are those of conventional masonry.

The decrease in the thermal amplitude of the T1 house in relation to the T2 house is graphically notable due to the thermal inertia of the high thermal mass, decreasing the air temperature by up to almost 3 °C compared to the lighter system in the afternoon when temperatures are at its peak. In other words, closed systems which use air conditioning during the day (such as business hours) would be the most benefited by using this construction system. However, the effect is reversed during the night, with the advantages and disadvantages according to the use having to be evaluated along with the permanence times of people in the structure. Figure 10 shows the results of monitoring with ventilation starting on February 14th and ending on February 26th.

It was possible to see the high decrease in the thermal amplitude in the system without ventilation, while in the system with ventilation the thermal inertia effect decreases a lot since there is heat being carried by the wind, either removing it from the system or adding it. In comparison with Fig. 10, the heat gain through ventilation shown in Fig. 9 in the T1 house is noticeable during the day, which previously managed to remain noticeably colder, but in this situation has a very similar behavior to T2, with a slight decrease in thermal amplitude, but the lines are already getting much closer.

With a much less pronounced curve than that perceived in Fig. 10, there is still great temperature variability in the afternoon with high standard deviations, which shows the potential of ventilation to interfere in the internal temperature of the environments. Even though it is understood that this experiment is done in order to enhance some parameters, it is clear that the ideal solution is more similar to Bioclimatic Zone 07 than could be perceived at first sight, as it involves selective ventilation during

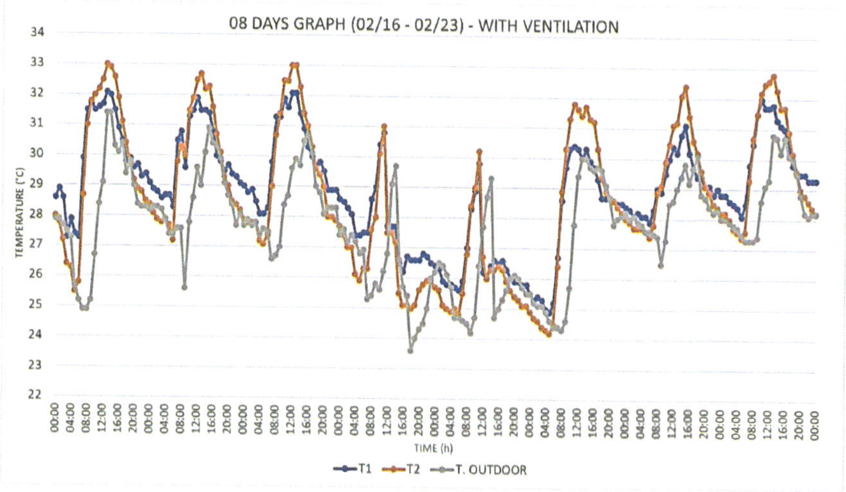

Fig. 10 Air temperature monitoring with ventilation

the day for this time of year, maintaining the mildest temperature on the walls, and protecting against heat gain from the outside.

In addition, the temperatures of the north–south walls of the two houses were measured in parallel to these measurements. Due to the sun positioning being a little further south in this period of the year, it was believed that they would have different temperatures due to the different solar charges. However, the experiment goes against this hypothesis, as highlighted in Figs. 11 and 12.

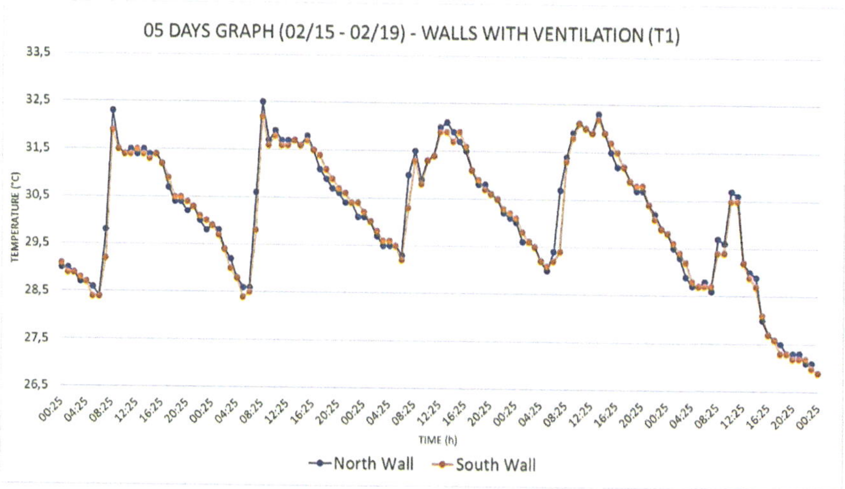

Fig. 11 Internal wall surface temperature (T1)

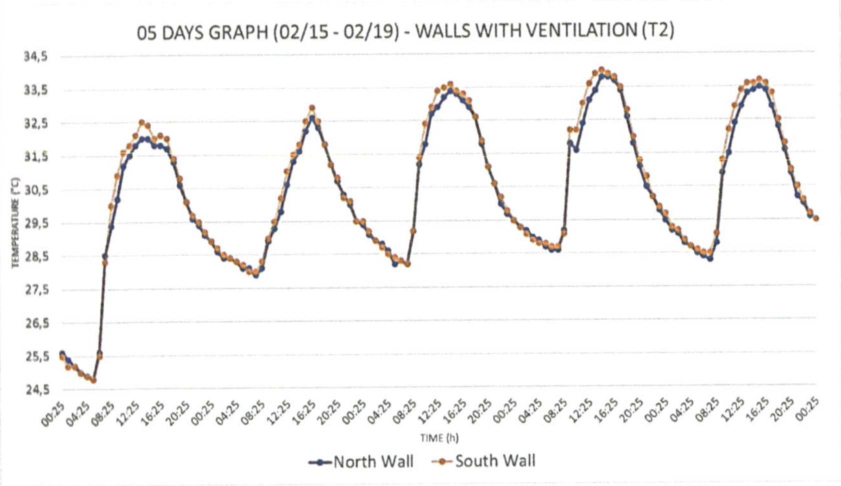

Fig. 12 Internal wall surface temperature (T2)

What stands out first when looking at the graphs is the pattern of the curves, in addition to the very similar values for opposite walls. While the T2 curves are almost perfect sine waves which explain the heat gain process, the T1 curves are more rigid with more prominent peaks, usually followed by a drop in temperature, possibly resulting from the heat flow paths going from warmer to colder and changing continuously. This permanent process is responsible for decreasing the thermal amplitude and is a consequence of the combined action of thermal delay and thermal inertia, in which the time it takes for the heat flow to go from the outside (hotter), to the inside (colder), lasts longer, and does not reach with intensity.

6.2 Acoustic Evaluation of Systems

At first, it was thought to take measurements on the spot with a decibel meter to obtain the actual sound transmission loss (STL), but there is a very busy route on the south facade of the terrain which shares the same wind direction, which is southeast. As the doors of the prototypes are facing east precisely to have the ventilation parameter in the thermal monitoring, the intermittent but frequent noise has a direct passage into the interior of the houses because they are only closed with a grid and the membrane used to prevent ventilation is light, thereby not providing good acoustic properties. This prevents a fair assessment to obtain the gross insulation values between rooms (*D*), defined by the difference of the equivalent sound pressure level (LA_{eq}) of the external environment, which was varying between 60 and 70 dB, while the internal environment (N_{EXT} and N_{INT}) did not decrease much, only about 5 dB. For this reason,

Table 12 Sound transmission loss (STL) in blocks and eight-hole bricks

CP	Area	Thickness (m)	d (kg/m^2)	M (kg/m^2)	STL (dB)	τ_i	Area × τ_i
SW_1	5	0.250	1571	392.77	64.88	0.000000	0.000002
SW_2	5	0.250	1665	416.20	65.39	0.000000	0.000001
SW_3	5	0.250	1582	395.56	64.94	0.000000	0.000002
SW_4	5	0.250	1570	392.59	64.88	0.000000	0.000002
SC_1	5	0.250	1872	467.97	66.40	0.000000	0.000001
SC_2	5	0.250	1910	477.42	66.58	0.000000	0.000001
SC_3	5	0.250	1849	462.37	66.30	0.000000	0.000001
SC_4	5	0.250	1909	477.34	66.58	0.000000	0.000001
CW_1	5	0.250	1850	462.59	66.30	0.000000	0.000001
CW_2	5	0.250	1870	467.57	66.40	0.000000	0.000001
CW_3	5	0.250	1740	435.03	65.77	0.000000	0.000001
CW_4	5	0.250	1758	439.38	65.86	0.000000	0.000001
CC_1	5	0.250	1714	428.45	65.64	0.000000	0.000001
CC_2	5	0.250	1736	433.91	65.75	0.000000	0.000001
CC_3	5	0.250	1707	426.79	65.60	0.000000	0.000001
CC_4	5	0.250	1735	433.81	65.75	0.000000	0.000001
Eight-hole brick	5	0.090	1150	103.50	53.30	0.000005	0.000023

sound pressure level measurements were not made due to the acoustic conditions of the terrain, but were calculated.

Therefore, the acoustic evaluation of the blocks took place numerically through Eq. (5) highlighted in the methodology, and the results can be seen in Table 12.

The difference between the two systems is easily noticeable, but it is an expected result considering the difference in mass and thickness. The difference of STL in block SC_2 (with better insulation) is higher than 13 dB for the brick due to this. However, even if the two systems were the same thickness, the masonry brick would still be less acoustically efficient due to the difference in density between the two. Both systems are within that established by the NBR 15575 [12] standard.

7 Conclusion

The results presented in this chapter were more positive than expected considering some reference texts from the literature review. However, they did not use cassava wastewater. The cassava wastewater used had a pH of around 4.5, considered acidic, and was predominantly formed by nitrogen, potassium, sodium, magnesium, and calcium according to a bench test. However, chemical analysis using X-ray fluorescence, which contains the composition in the form of oxides, showed a predominance of calcium (CaO), iron (Fe_2O_3), potassium (K_2O), and silica (SiO_2). For the soil, mineralogical analysis revealed that its predominant clay was kaolinite, followed by quartz and illite. The chemical analysis presented a high concentration of iron (Fe_2O_3), followed by silica (SiO_2) and alumina (Al_2O_3), and its characterization presented consistent values with those established by the standards for soil-cement, which is the main reference of this work due to the lack of specific standards.

In analyzing the composition and defining the optimum moisture in 10% for the system through the compaction test, which presented more favorable results for the soil and the cassava wastewater, a total of four formulations were established to submit compression strength, water absorption by capillarity, and thermal conductivity tests: soil and water (SW); soil and cassava wastewater (SC); soil, water, and 10% cement (CW); and soil, cassava wastewater, and cement (CC). All compositions obtained strengths higher than 1 MPa, which is the minimum required for sealing in soil-cement, and with the exception of SW blocks, all others reached more than 2 MPa. The cement compositions showed less thermal conductivity than those which were not stabilized, and it is believed that this is due to the permeability difference in which the blocks with only soil and cassava wastewater got denser, thus facilitating thermal conduction.

From the thermal evaluation of the prototypes, it was noticeable that, when in the absence of ventilation, T1 has a considerable decrease in the thermal amplitude, with lower temperature values during the day, but higher during the night. When the system is ventilated, T1 behavior is similar to T2, considering that ventilation during the day also brings heat into space, making it warmer than if this ventilation was not allowed to enter. Based on this, it is highlighted that the ideal use of the system would be: ventilating during the night and avoiding ventilation during the day. In acoustic terms, due to the mass law, the denser the material, higher its insulating capacity, like the one presented by the soil-cassava wastewater composition. Thus, blocks with cassava wastewater and with no cement showed similar results to the ones with it, showing the possibilities with the proposed system.

Acknowledgements This study was financed in part by the *Coordenacao de Aperfeicoamento de Pessoal de Nivel Superior*—Brazil (CAPES)—Finance code 001, the Federal Institute of Education, Science and Technology of Paraiba and the Federal University of Rio Grande do Norte.

References

1. Brazil (1978) Ministry of Labour. NR-15-unhealthy activities and operations - annex 11
2. Brasil (2005) Instruction booklet CI 10-7/1: subsistence logistics. Terrestrial, E. B. C. D. O. Defense Ministry, Brasília
3. Brazilian Association of Technical Standards—ABNT (2016a) NBR 6459, Soil – determination of the liquid limit. Rio de Janeiro

4. Brazilian Association of Technical Standards—ABNT (2016b) NBR 7180, Soil – determination of the plastic limit. Rio de Janeiro
5. Brazilian Association of Technical Standards—ABNT (2016c) NBR 7181, Soil—granulometric analysis. Rio de Janeiro
6. Brazilian Association of Technical Standards—ABNT (2016d) NBR 7182, Soil - compaction test. Rio de Janeiro
7. Brazilian Association of Technical Standards—ABNT (2012a) NBR 8492, Soil-cement brick - dimensional analysis, determination of compressive strength and water absorption - test method. Rio de Janeiro
8. Brazilian Association of Technical Standards—ABNT (2012b) NBR 10833, manufacture of brick and block of soil-cement using manual or hydraulic press - procedure. Rio de Janeiro
9. Brazilian Association of Technical Standards—ABNT (2012c) NBR 13553, materials for use in monolithic soil-cement wall without structural function - requirements. Rio de Janeiro
10. Brazilian Association of Technical Standards—ABNT (2005a) NBR 15220, thermal performance of buildings - Parts 01, 02, 03, 04 and 05. Rio de Janeiro
11. Brazilian Association of Technical Standards—ABNT (2005b) NBR 15259, mortar for laying and covering walls and ceilings - determination of water absorption by capillarity and capillarity coefficient. Rio de Janeiro
12. Brazilian Association of Technical Standards—ABNT (2013) Brazilian standard NBR 15575, residential buildings - performance. Rio de Janeiro
13. Cardoso CLE, Souza JS, Gameiro AH (2006) Economic aspects and markets. In: Socioeconomic and agronomic aspects of cassava. empraba: cassava and fruit. Cruz das Almas, Brazil
14. Castro APAS, Labaki LC, Caram RM, Basso A, Fernandes MR (2003) Solar reflectance of colours for external paints obtainedby spectrophotometric analysis. Ambiente Construído. Porto Alegre 3(2):69–76
15. CBCS (2007) About CBCS. Available at: http://www.cbcs.org.br/website/. Accessed on Dec 2019
16. CBCS (2013) Action guidelines - Review 1. Available at: http://www.cbcs.org.br/website/. Accessed on Dec 2019
17. de Souza JM (2019) Soil-cement bricks produced with cassava wastewater instead of water. Master dissertation. Federal University of Rio Grande do Norte, Technology Center, Graduate Program in Civil Engineering, Natal/RN, 80 p
18. dos Santos JD, Veit MT, Palácio SM et al (2017) Evaluation of the combined process of coagulation/flocculation and microfiltration of cassava starch wastewater: removal efficiency and membrane fouling. Water Air Soil Pollut 228:238
19. Dutra S, Vale C (2016) The use of lateritic soils as road pavement material. Geotecnia 136:81–93
20. Jornal Nacional (2010) Mud houses are a health threat. Available at: http://g1.globo.com/jornal-nacional/noticia/2010/04/casas-de-taipa-sao-uma-ameaca-saude.html. Accessed on Dec 2019
21. Khalili EN (1989) Lunar structures generated and shielded with on-site materials. J Aerosp Eng 2(3):119–129
22. Minke G (2001) Construction manual for earthquake-resistant houses. Forschungslabor für Experimentelles Bauen. University of Kassel, Germany
23. Pedrosa MS, Xavier T (2017) Brickeradobe: methodological proposal for application in rural and indigenous areas. Research Report. University Center of Brasilia, Brazil
24. Peres S, Monteiro MR, Ferreira ML et al (2019) Anaerobic digestion process for the production of biogas from cassava and sewage treatment plant sludge in Brazil. BioEnergy Res 12:150–157
25. Pisani MAJ, Correa P, Caldana V et al (2006) Experimental site: practice or invention? Aedificandi Revista de Arquitetura e Construção (Especial). Sao Paulo I(2):1–15
26. Rodriguez DA, Stein J, Darby T (2018) Cyclic testing of reinforced earthbag walls. In: Civil Engineering Theses, vol 68, Santa Clara University
27. Santos CA (2015) Earth construction in Brazil: overview, standardization and prototyping with earthbag. Master dissertation. Universidade Federal de Santa Catarina, Florianópolis/SC, Brazil, 290 p

28. Seco A, Urmeneta P, Prieto E et al (2017) Estimated and real durability of unfired clay bricks: Determining factors and representativeness of the laboratory tests. Constr Build Mater 131:600–605
29. Silva FF, Bertonha A, Freitas PSL et al (2005) Fragment of starch to glucose to identify waste water from manioc industry in depth of soil. Acta Sci Biol Sci 27:507–511
30. Souza RS (2019) Mortar produced from the substitution of the fine aggregate by residue of the scheelite and the water of hydration by cassava wastewater. Federal University of Rio Grande do Norte, Technology Center, Graduate Program in Civil Engineering, Natal/RN, 127 p
31. Tokarnia CH (2012) Toxic plants from Brazil for farm animals. Helianthus, Rio de Janeiro, Brazil, pp 443–459
32. Velde B (1992) Introduction to clay minerals. chemistry, origins, uses and environmental significance. London, Chapman & Hall, England
33. Villibor DF, Nogami JS, Cincerre JR, Serra RMS, Zuppolini Neto A (2009) Low-cost pavements for urban roads, 2nd edn. Arte e Ciencia, Sao Paulo, Brazil
34. Zimmer T, Souza VA, Romio LC, Buligon L, Roberti DR, Moreira VS (2018) Study of the thermal conductivity of the soil concerning the moisture content in a pasture area in the pampa biome. In: X Workshop Brasileiro de Micrometeorologia, Santa Maria

The manufacturer's authorised representative in the EU is Springer Nature Customer Service Centre GmbH, Europaplatz 3, 69115 Heidelberg, Germany. If you have any concerns regarding our products, please contact ProductSafety@springernature.com

Printed and bound by CPI Group (UK) Ltd, Croydon, CR0 4YY
30/09/2026
02225665-0002